FORBIDDEN CONFIGURATIONS IN DISCRETE GEOMETRY

This book surveys the mathematical and computational properties of finite sets of points in the plane, covering recent breakthroughs on important problems in discrete geometry and listing many open problems. It unifies these mathematical and computational views using forbidden configurations, which are patterns that cannot appear in sets with a given property, and explores the implications of this unified view. Written with minimal prerequisites and featuring plenty of figures, this engaging book will be of interest to undergraduate students and researchers in mathematics and computer science.

Most topics are introduced with a related puzzle or brain-teaser. The topics range from abstract issues of collinearity, convexity, and general position to more applied areas including robust statistical estimation and network visualization, with connections to related areas of mathematics including number theory, graph theory, and the theory of permutation patterns. Pseudocode is included for many algorithms that compute properties of point sets.

David Eppstein is Chancellor's Professor of Computer Science at the University of California, Irvine. He has more than 350 publications on subjects including discrete and computational geometry, graph theory, graph algorithms, data structures, robust statistics, social network analysis and visualization, mesh generation, biosequence comparison, exponential algorithms, and recreational mathematics. He has been the moderator for data structures and algorithms on arXiv.org since 2006 and is a major contributor to Wikipedia's articles on mathematics and theoretical computer science. He was elected as an ACM fellow in 2012.

Forbidden Configurations in Discrete Geometry

DAVID EPPSTEIN
University of California, Irvine

CAMBRIDGE
UNIVERSITY PRESS

CAMBRIDGE
UNIVERSITY PRESS

University Printing House, Cambridge CB2 8BS, United Kingdom

One Liberty Plaza, 20th Floor, New York, NY 10006, USA

477 Williamstown Road, Port Melbourne, VIC 3207, Australia

314–321, 3rd Floor, Plot 3, Splendor Forum, Jasola District Centre, New Delhi - 110025, India

79 Anson Road, #06-04/06, Singapore 079906

Cambridge University Press is part of the University of Cambridge.

It furthers the University's mission by disseminating knowledge in the pursuit of education, learning, and research at the highest international levels of excellence.

www.cambridge.org
Information on this title: www.cambridge.org/9781108423915
DOI: 10.1017/9781108529167

First published 2018

Printed in the United States of America by Sheridan Books, Inc.

A catalogue record for this publication is available from the British Library

ISBN 978-1-108-42391-5 Hardback

Contents

Acknowledgments

This book stemmed from an invitation to present the Erdős Memorial Lecture at the 29th Canadian Conference on Computational Geometry (held in Ottawa in July 2017) and would not have existed without that invitation. I would like to thank the many people who have given me helpful advice on it, especially Jean Cardinal, Sariel Har-Peled, Stefan Langerman, Joe O'Rourke, János Pach, Vijay Vazirani, and several anonymous reviewers. I am also grateful for the careful copyediting of Maureen Eppstein. The research in this work was supported in part by the US National Science Foundation under grants CCF-1618301 and CCF-1616248.

1 A Happy Ending

In the early 1930s, Hungarian mathematician Esther Klein made a discovery that, despite its apparent simplicity, would kick off two major lines of research in mathematics. Klein observed that every set of five points in the plane has either three points in a line or four points in a convex quadrilateral. This became one of the first results in the two fields of discrete geometry (the study of combinatorial properties of geometric objects such as points in the Euclidean plane, and the subject of this book) and Ramsey theory (the study of the phenomenon that unstructured mathematical systems often contain highly structured subsystems).

Klein's observation can be proven by a simple case analysis that considers how many of the points belong to their *convex hull*. The convex hull is a convex polygon, having some of the given points as its vertices and containing the others. It can be defined mathematically in many ways, for instance as the smallest-area convex polygon that contains all of the given points or as the largest-area simple polygon whose vertices all belong to the given points. The convex hull of points that are not all on a line always has at least three vertices (for otherwise it could not enclose a nonzero area) and, for five given points, at most five vertices. If it has five vertices, any four of them form a convex quadrilateral, and if it has four vertices then it is a convex quadrilateral. The remaining possibility for the convex hull is a triangle, with the other two points either part of a line of three points or inside the triangle. When both points are inside, and the line through them misses the triangle vertices, it also misses one side of the triangle. In this case the two interior points and the two points on the missed side form a convex quadrilateral (Figure 1.1).

The challenge of extending and generalizing this observation was taken up by two of Klein's friends, Paul Erdős and George Szekeres. They proved that,

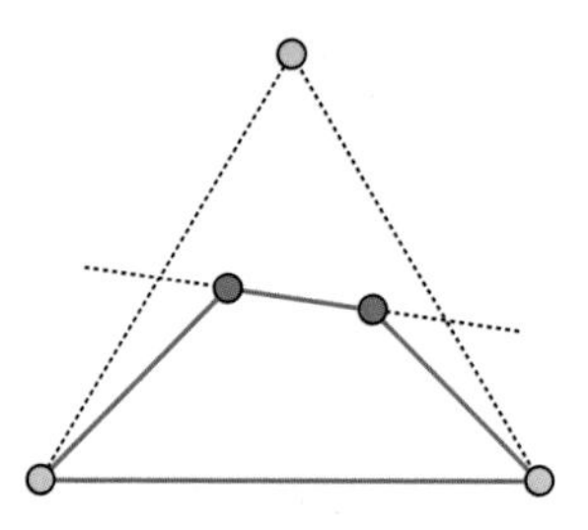

Figure 1.1. Five points with no three in line always contain a convex quadrilateral, either four vertices of the convex hull or (as shown here) two vertices of a triangular convex hull and two interior vertices. When the convex hull is a triangle, the line through the remaining two points has two triangle vertices on one of its sides, and those four points are in convex position.

for every k, a convex k-gon can be found in all large enough sets of points, as long as no three of the points lie on a line.[1] Klein later married Szekeres, and their marriage is commemorated in the name of Erdős and Szekeres's result: the happy ending theorem.

Erdős and Szekeres published two proofs of their theorem, one of which showed that every 4^k points in general position (meaning, no three in a line) contain a convex k-gon.[2] However, this gives only a loose estimate. For instance, it would tell us that we need $4^4 = 256$ points to guarantee the existence of a convex quadrilateral, many more than the five points of Klein's observation. Therefore, it became of interest to determine more precisely how many points are needed to ensure the existence of a convex k-gon.

In their original work on this problem, Erdős and Szekeres conjectured that many fewer points, $2^{k-2}+1$ of them, would already force a convex k-gon to exist. Later, they constructed sets of 2^{k-2} points with no three in line and no convex k-gon, so if true their conjecture would be as tight as possible.[3] For example, some sets of eight points have no convex pentagon, matching the formula as $8 = 2^{5-2}$ (Figure 1.2). We detail their construction in Section 11.1. Tightening the gap between this construction and the 4^k upper bound remained open until in a recent breakthrough Suk (2017) proved that for sufficiently large k, every $2^{k+6k^{2/3}\log k}$ points in general position contain a convex k-gon. Although this does not settle the conjecture of Erdős and Szekeres, it has the correct leading term in the exponent and brings the upper and lower bounds much closer.[4]

[1] Erdős and Szekeres (1935).

[2] More precisely, they showed that every set of $\binom{2k-4}{k-2}+1 \leq 4^k$ points has a convex k-gon. Here $\binom{2k-4}{k-2}$ is a *binomial coefficient*, the number of ways of choosing $k-2$ elements from a set of $2k-4$ elements.

[3] Erdős and Szekeres (1960).

[4] Suk writes that Gábor Tardos has further improved the low-order term in the exponent of this bound. For an intuitive overview of Suk's proof, see Hartnett (2017).

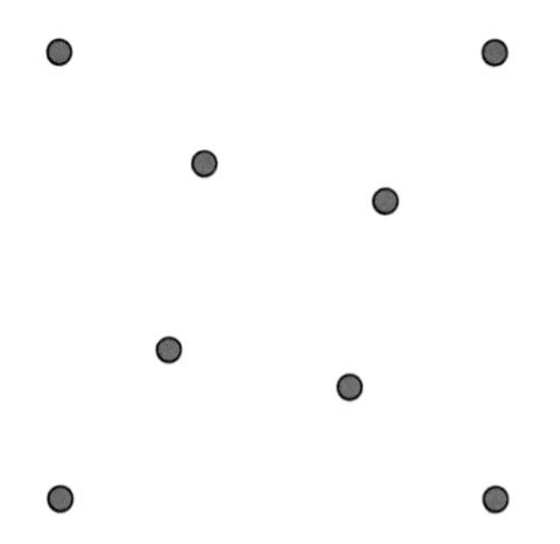

Figure 1.2. Eight points in general position that do not contain the vertices of a convex pentagon.

Open Problem 1.1 (the happy ending problem)

Does there exist an integer k, and a set of $2^{k-2}+1$ points in the plane with no three in a line and no k forming a convex k-gon?

Two features of the happy ending theorem, and of the convex subsets of point sets that it describes, are of particular interest to us. First, the size of the largest convex subset of a set of points is *monotone*: if you remove points from the set, then its largest convex subset can only decrease in size or stay the same, but it can never grow. Second, the convex subsets of point sets are insensitive to the precise locations of the points. If you move the points around the plane in a continuous motion, being careful only to never let three of them line up, you cannot create new convex polygons nor destroy the ones that are already present. Another way to express the same insensitivity is that the convex subsets of a point set depend only on the *orientations* of the points: which triples are in clockwise order, which counterclockwise, and which collinear. Many other problems in discrete geometry share these characteristics: they involve monotone properties of finite point sets that depend only on the orientations of the points. Problems of this type are the subject of this book.

2 Overview

Many algorithmic and combinatorial problems concerning finite sets of points have been studied in discrete and computational geometry. Often, the answers to these problems depend only on knowing, for each three points, whether they are in clockwise order, counterclockwise order, or lie on a single line. It is safe, for these problems, to throw away the coordinates of the points and retain only their *configuration*, which tells us this ordering information for each triple of points. In many cases, in addition, the property or quantity to be studied behaves predictably under the removal of points. If removing a point can never cause a quantity of interest to increase, we call that quantity *monotonic*. Our goal in this work is to provide a systematic study of the monotonic properties of configurations.

Several old and colorfully named puzzles and games fit this pattern:

- The happy ending theorem was famously given its name after its proof led to the marriage of two of the mathematicians who discovered it, Esther Klein and George Szekeres. It is about how many points are needed (no three in a line) to ensure the existence of a convex polygon with a given number of corners. We described it already in Chapter 1.
- The orchard-planting problem, which we describe in Section 8.1, dates back to the early nineteenth century. It asks how many rows of three trees one can form by planting an orchard with a given number of trees.
- We describe the no-three-in-line problem in Section 9.1. It was first posed in terms of placing 16 pawns on a chessboard, so that no three of them line up with each other. For this problem, it is important to consider all directions of lines, not just the horizontal, vertical, and diagonal directions of the chessboard.

Other topics of more serious past research also concern monotonic properties of configurations. They include searching for triples of points that all lie on a single line (Chapter 8), grouping points into clusters of collinear points (also Chapter 8), finding convex polygons within sets of points (Chapter 11), partitioning sets of points into nested convex polygons ("onion layers," Section 12.4), estimating the center of a cloud of points in a way that is insensitive to perturbations of the points (Section 12.7), perturbing sets of points so their distances or coordinates are all integers (Chapter 13), using sets of points as vertices to draw planar graphs (Chapter 16), and finding paths that no line can cross many times (Section 17.1).

Our study of the monotone properties and parameters of configurations looks at them from the following points of view.

Forbidden configurations. Each monotone property, and each value of a monotone parameter, can be characterized by its *forbidden configurations* or *obstacles*. These are the configurations that do not have the property (or that have too large a value) but for which all subconfigurations do have it. The properties of any particular configuration can be read off from whether it contains any of these obstacles.

For the properties and parameters we consider, we ask: are there a finite number of forbidden configurations? If so, can we describe them all, or bound their size as a function of k?

Computational complexity. Can the given property or parameter be computed in polynomial time, or is it NP-hard? If a parameter is hard, how well can it be approximated, and how efficiently can we compute small parameter values? Can we distinguish point sets with a property from sets far from having the property by examining only small samples of the set, or is it necessary to test the whole input?

Both the problems of computing small parameter values and of using samples to test properties are closely related to the existence of finitely many obstacles. As we will see, algorithms using a technique called *kernelization* can often be used to bound the size of the obstacles for a parameter. And the size of a sample needed to test any property can be bounded in terms of the size of the obstacles for the property. However, we will also see that in some cases a parameter may have a constant number of obstacles for each parameter value but still be hard to compute, even for bounded parameter values.[1]

Inequalities. How are different monotone parameters related to each other? Which ones are bounded above or below by functions of each other?

Well-quasi-ordering. For which families of configurations do all monotone properties on that family have a finite number of obstacles? The families

[1] For the connection to sampling, see Section 6.5. For a parameter that is hard for bounded parameter values, see Theorem 7.7.

for which this is true are called *well-quasi-ordered*. The family of all configurations is not well-quasi-ordered, but some of its subfamilies are, and we investigate which ones.

Combinatorial enumeration. How many configurations have a given property? The number of all n-point configurations is known to grow exponentially in $n \log n$.[2] Is the growth rate of the configurations with a given property or obstacle significantly smaller?

Along the way we will also collect a large menagerie of monotone properties and monotone parameters of configurations. In addition we investigate the complexity of testing whether one configuration is part of another larger one. This can be done quickly when the size of the smaller configuration is fixed, but is much harder when its size can vary.

Although much of this work should be readable by nonspecialists, some of it remains technical. Most chapters place the more generally accessible aspects of their subject in the earlier parts of the chapter, and the more technical aspects later, so readers who find some material difficult should feel free to skip ahead to the start of the next chapter where it will likely be easier going again. It is generally safe to skip past proofs, at least on a first reading, and many of the results in the earlier chapters of the book have proofs that we have delayed until the later chapters. Chapters 3–5 give the main definitions that we use, of configurations, subconfigurations, monotone parameters, and monotone properties, and some important general results based on these definitions. Chapters 6 and 7 introduce the algorithmic study of configurations. The remaining chapters are largely independent from each other. They discuss different subtopics of discrete geometry that all involve monotone parameters of configurations.

Much of this work has analogies with the theory of graphs and subgraphs (or graphs and minors) and with the theory of permutations and permutation patterns. For instance, the questions raised above about combinatorial enumeration are analogous to the Stanley–Wilf conjecture for permutations, proved by Marcus and Tardos (2004), according to which forbidding a single permutation pattern reduces the number of permutations in a permutation class from factorial to single exponential. We will exploit the analogy to permutations and permutation patterns in Chapter 14 by finding general methods of translating permutations into configurations. Similarly in Chapter 15 we translate graphs into configurations. We use these translations both positively, to develop efficient algorithms for configurations, and negatively, to show that certain problems on configurations are hard to compute.

We conclude with Chapter 18, which summarizes the results from these chapters and provides a road map of the relations between parameters from different chapters.

[2] Goodman and Pollack (1986).

Because of our focus on monotone properties of configurations, there are many aspects of discrete geometry that are beyond the scope of our work. For instance, there have been recent breakthroughs on the number of distinct distances that any point set in the plane must have,[3] and on related problems of counting incidences of geometric objects, that we do not cover here.

[3] Guth and Katz (2015).

3 Configurations

We begin our study with an examination of the different ways we can place small numbers of points in the plane, and what it means for two sets of points to be different.

3.1 Small Configurations

In how many different ways can we place n points in the plane? With a list of all of the possible placements, we could prove statements such as Klein's observation about convex quadrilaterals in five-point sets, automatically, merely by checking all the cases. When n is small enough, we can provide an explicit answer.

Example 3.1

There are two different ways of placing three points in the plane: they may either lie on a line, or they may form a triangle.

Four points may be arranged in four different ways:

- a four-point line
- three points on a line and one off the line
- a triangle containing one point, or
- a convex quadrilateral.

Figure 3.1 depicts these three-point and four-point configurations.

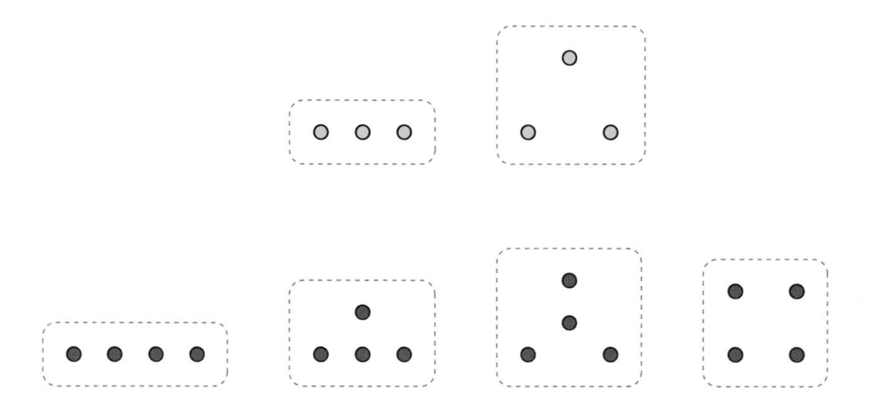

Figure 3.1. Two ways of arranging three points (yellow, top) and four ways of arranging four points configurations (red, bottom).

So far, we have used only an intuitive notion for what it means for two sets of points to be the same or different. But when we get to five points, we already need to define more precisely what we mean. Figure 3.2 depicts 13 sets of five points. Are they all different from each other? Two of these are mirror images: should that count as two different sets of points or as two different views of a single way of placing five points?

We will give a more precise definition of what it means for sets of points to be the same or different in the next section. For the definition we use, mirror images are not (in general) considered to be the same, so there are indeed 13

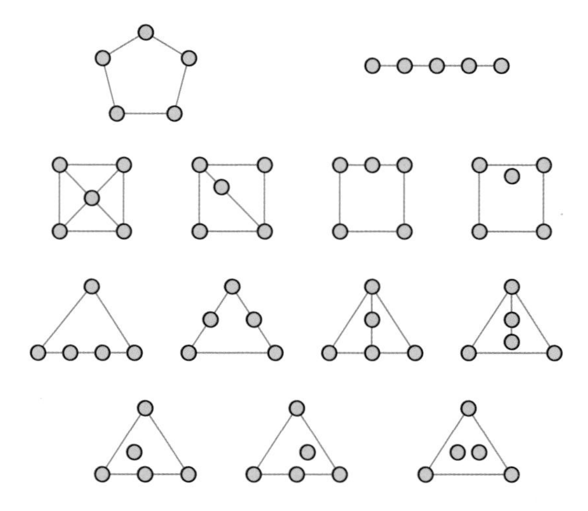

Figure 3.2. The 13 possible five-point configurations. Each configuration is shown together with its convex hull edges and with any line segments that pass through three or more collinear points.

different ways of arranging five points. Open Problem 3.5, later in this chapter, formalizes the problem of counting configurations of different sizes.

The numbers of sets of n points grow very quickly, as $c^{n \log n}$ for a constant $c > 1$.[1] Aichholzer et al. (2002) used computer searches to establish a database of small sets of points.[2] To help control the size of the database, they exclude sets that have three points on a line and count mirror-image sets of points as equivalent. Despite these restrictions, the sets in their database still grow as $c^{n \log n}$, but with a smaller c. Their numbers are:

n:	1	2	3	4	5	6	7	8	9	10	...
#:	1	1	1	2	3	16	135	3315	158817	14309547	...

For instance the two configurations they count with $n = 4$ are the two on the lower right of Figure 3.1.

3.2 Orientations and Order Types

To test whether two sets of points are placed in the same way or differently, we use *order types*. Intuitively, the order type of a set of points in the plane describes how the points are positioned with respect to each other: which points are on lines with each other, and how do the lines through pairs of points split the remaining points? We define these concepts more formally below. Our definitions will allow us to define the shape of a set of points in an abstract way that does not depend on how the points are scaled or rotated or on other inessential properties.

The building blocks of order types are the orientations of triples of points.[3]

Definition 3.2

We define the *orientation* of an ordered triple of points in the Euclidean plane to be one of the three numbers $+1$, -1, or 0. It is $+1$ if the three points form the vertices of a nonzero-area triangle, listed by the triple in clockwise order around the triangle. It is -1 if they are in counterclockwise order, or 0 if they are collinear.

Two cyclic permutations of the same triple have the same orientation, and reversing a triple causes its orientation to be negated. Therefore, if we know the

[1] Goodman and Pollack (1986).

[2] This database is online at www.ist.tugraz.at/aichholzer/research/rp/triangulations/ordertypes/.

[3] It is not possible to break down this information further, into locally defined structures on pairs of points; see Balko et al. (2017).

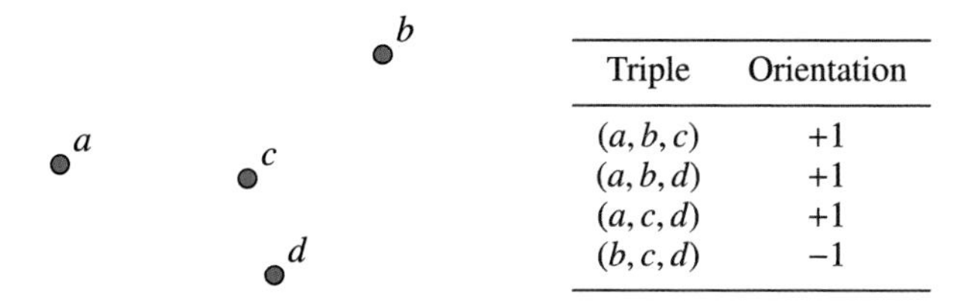

Triple	Orientation
(a,b,c)	$+1$
(a,b,d)	$+1$
(a,c,d)	$+1$
(b,c,d)	-1

Figure 3.3. The order type of a set of four points (Definition 3.3)

orientation of a single ordered triple, we also know the orientation of its other five permutations.

Definition 3.3

The *order type* of a set of points in the Euclidean plane is the mapping from ordered triples of points to their orientations.

See Figure 3.3 for an example. The orientations of the 24 triples of these four points can be recovered from permutations of the four triples whose orientations are listed in the table.

Not every mapping from triples of objects to the numbers $+1$, -1, and 0 can be an order type of a set of points. The order types obey certain constraints, some of which can be formalized using abstract mathematical structures called oriented matroids, while others are specific to Euclidean point sets.[4] However, we will not use this theory explicitly here.

Many authors in discrete and computational geometry restrict their arguments to sets of points that are in *general position*. What they mean can vary, but the basic idea is that moving points by small distances should not change the combinatorial behavior of the problem. This restriction simplifies many arguments, and some general tools can automatically extend results from points in general position to other sets of points.[5] For the problems we consider, what can change when we move points is whether they are collinear. Moving three points from starting positions on a line would, in general, cause them to stop lining up. Therefore, we will say that points are in general position when no more than two points lie on any line, or equivalently when there are no zeros in their order type. However, although this will be an important concept for us, we will usually avoid assuming that points are in general position. In fact, many problems that we consider make sense only for sets of points that include lines of three or more points.

[4] Knuth (1992); Björner et al. (1999). [5] Edelsbrunner and Mücke (1988).

3.3 Defining Configurations

The next definition is of configurations, the primary objects that we study.

Definition 3.4

Two finite sets of points S and T are *order-equivalent* if we can map the points of S one-to-one onto the points of T in a way that preserves the orientation of each triple of points. We define the *configuration* of a point set S to be the class of all finite sets of points that are order-equivalent to S. More generally, we define a configuration to be any class of finite sets of points that can be described in this way.[6]

We define a *realization* of a configuration S to be one of the finite sets of points that belongs to S. When we talk about the points of a configuration, we will use this as a shorthand for the points of an arbitrarily chosen realization of the configuration.

With this definition in hand we can formalize the counting problem described in Section 3.1.

Open Problem 3.5

How many configurations of n points are there, as a function of n? Is there a formula for this number?

Order-equivalence is an *equivalence relation*, meaning that it obeys the following three axioms:

- Reflexivity: Every point set is order-equivalent to itself.
- Symmetry: For every two point sets X and Y, if X is order-equivalent to Y, then Y is order-equivalent to X.
- Transitivity: For every three point sets X, Y, and Z, if X is order-equivalent to Y and Y is order-equivalent to Z, then X is order-equivalent to Z.

Whenever we have an equivalence relation, we can partition the things that it relates into *equivalence classes*, collections of objects that are all equivalent

[6] Our definition of a configuration differs from that of Grünbaum (2003), for instance. Grünbaum's definition considers only whether two sets of points have the same collinearities, and not whether they agree on the orientations of their noncollinear triples.

to each other. The configurations, as defined above, are just the equivalence classes for order-equivalence.

It is convenient to have a shorthand notation for the number of points in a configuration.

Definition 3.6

We define the *size* $|S|$ of a configuration S to be the number of points in any set of points that represents it.

This definition does not depend on the choice of a set of points to represent a given configuration: all of these sets have the same number of points.

3.4 Example Configurations

In this section, we give names to some familiar examples of configurations in which the points form straight lines, convex polygons, and grids. As with most of the notation that we will use for configurations, their properties, and parameters, we will spell out names for these configurations as words in small-capital letters, rather than relying on a plethora of hard-to-remember single-letter notations in obscure alphabets.[7]

Definition 3.7

We define $\textsc{grid}(n, m)$ to be the configuration of an $n \times m$ rectangle of integer points,

$$\{(x, y) \mid x, y \in \mathbb{Z} \text{ and } 0 \leq x < n \text{ and } 0 \leq y < m\}.$$

As a special case, we define $\textsc{line}(n)$ to be $\textsc{grid}(n, 1)$, the configuration of n collinear points. Any n points on a single line are order-equivalent to $\textsc{line}(n)$. Their order type is independent of their spacing or positions along the line.

Definition 3.8

$\textsc{polygon}(n)$ is the configuration of vertices of a convex n-gon.

[7] Computer programmers have long known that mnemonic variable names are easier to understand (Shneiderman and Mayer, 1979). However, this lesson does not seem to have reached much of mathematics nor even the more mathematical parts of computer science.

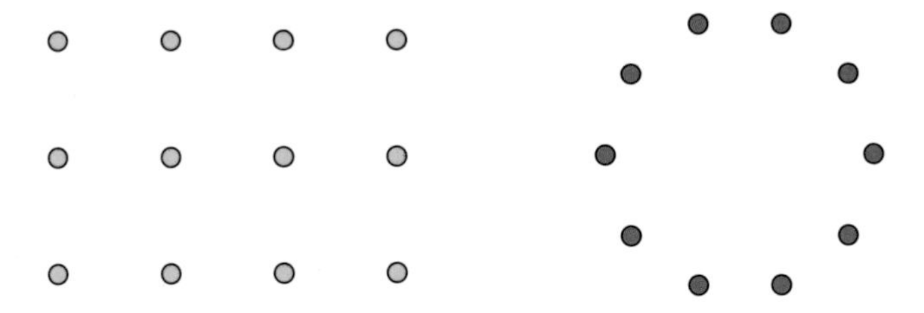

Figure 3.4. Left: GRID$(4, 3)$ (Definition 3.7). Right: POLYGON(10) (Definition 3.8).

All convex n-gons have order-equivalent sets of vertices: their order-type cannot distinguish regular polygons (in which all side lengths and all vertex angles are equal) from other convex polygons. Nevertheless, it is helpful to think of POLYGON(n) as being represented by a regular n-gon.

The configurations GRID$(4, 3)$ and POLYGON(10) are illustrated in Figure 3.4. The next set of examples is less familiar, but has convenient properties that we will use later.

Definition 3.9

We define SAWTOOTH(n) to be the configuration of a set of $2n$ points, n of which form the vertices of a convex n-gon, and the other n of which are interior to the n-gon, near the midpoint of each edge of the n-gon. Here, by "near" we mean that there is no line through two points of the configuration that separates any of these interior points from the edge midpoint that it is near.

The name reflects the resemblance of these configurations to circular saw blades. Examples of the sawtooth configurations are shown in Figure 3.5.

3.5 Partitions by Lines

The order type does not directly describe how the points within a line are ordered. This sort of ordering could be axiomatized, for instance, by using the

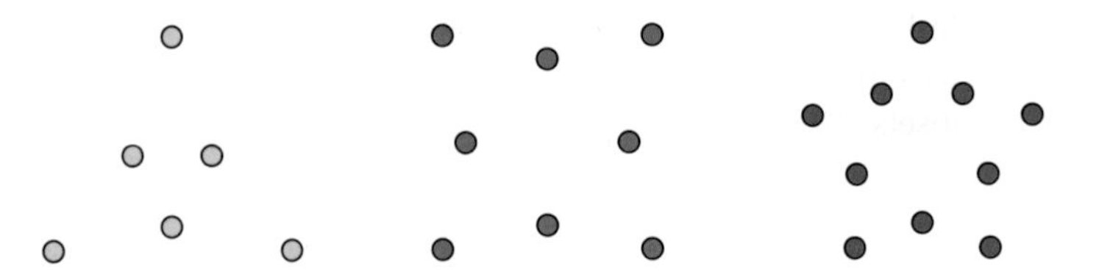

Figure 3.5. The configurations SAWTOOTH(3) (left, yellow), SAWTOOTH(4) (middle blue), and SAWTOOTH(5) (right, red); see Definition 3.9.

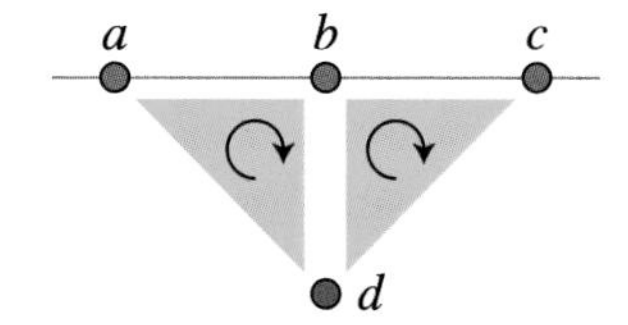

Figure 3.6. Notation for the proof of Lemma 3.10: b is the middle of three collinear points a, b, and c if and only if, for any point d not on the line, (a, b, d) and (b, c, d) have the same orientation.

betweenness relation on triples of points.[8] But although we do not represent betweenness explicitly, the order of the points on a line can also be recovered from the order type, in most cases, as the following lemma shows.

Lemma 3.10

Let C be a configuration in which the points of C do not lie on a single line. Then, for every line L through three or more points of C, the ordering of points along L is determined by the order type of C.

Proof. Suppose we wish to identify the middle of three collinear points a, b, and c. Let d be an arbitrary point that is not on the same line as these three points (Figure 3.6). Then the middle point is b if and only if the two ordered triples (a, b, d) and (b, c, d) have the same orientation. □

As we now show, the order type of a set of points is also enough to determine how it can be partitioned by lines.

Definition 3.11

We define a *line partition* of a set of points S to be a partition of S determined by a line L that does not pass through any point of S. It splits S into two subsets, one on each side of L. We view these subsets as unordered: swapping them gives the same partition. We define an *anchored line partition* to be another kind of partition of S, from an ordered pair (p, q) of distinct points of S. It places the points r for which (p, q, r) has orientation $+1$ into one of the two subsets. Every point r for which (p, q, r) has orientation 0 or -1 goes, with p and q, into the other subset.

[8] Chor and Sudan (1998).

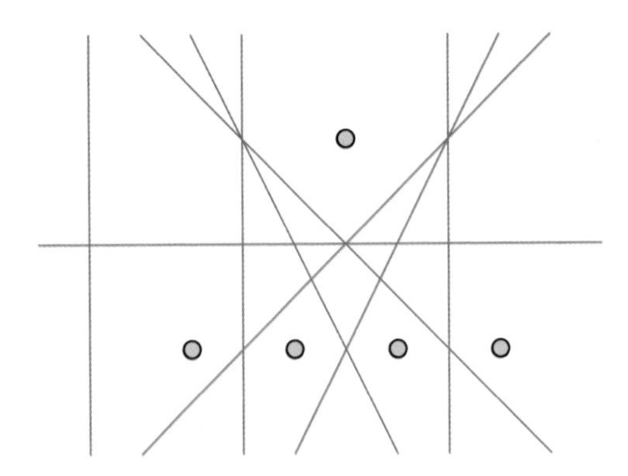

Figure 3.7. A set of five points with eight different line partitions.

Figure 3.7 depicts the eight line partitions of a five-point set (including a partition with one side empty and the other side including all the points). Anchored line partitions are determined only by the order type of S, because they use only orientations in their definition. Every anchored line partition is a line partition, for a line parallel to line pq but perturbed to one side of it.

Lemma 3.12

For every finite set of points S that are not all on a single line, every line partition of S is an anchored line partition.

Proof. Let L be a line that forms a line partition of S. We need to find two points p and q such that the same partition is an anchored line partition for p and q. We will assume that both sides of the partition are nonempty, for otherwise the result is easy: just choose p and q to be the endpoints of a convex hull edge.

Let A and B be the convex hulls of the subsets of S on either side of L, as shown in Figure 3.8. Like any two disjoint convex polygons, A and B are separated by two *bitangents,* lines that touch both A and B at a vertex without crossing either polygon (the red lines of the figure). (Another two bitangents do not separate A from B, but they might coincide with the separating bitangents when A or B is a single point.) The assumption that S is not all on one line causes these two separating bitangents to be distinct from each other, because all of the points of S are contained within the double wedge formed by the two bitangents, and this double wedge must have nonzero area.

At most one of A and B can have a point at the crossing of the two bitangents. Suppose by symmetry that (as in the figure) A is separated from this crossing point. Because of this separation, the bitangents touch A at two points p and r. These two points split the boundary of A into two convex chains, an inner one closer to the crossing point of the bitangents and an outer one farther from

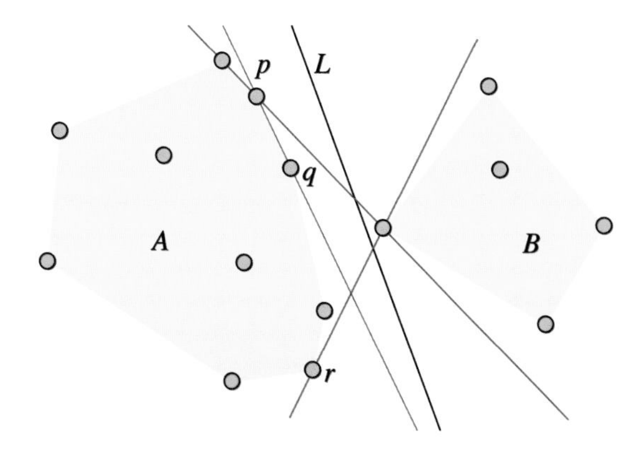

Figure 3.8. Illustration for the proof of Lemma 3.12. Given a line partition by a line L (black), we find the convex hulls A and B of the points on either side of the line (shaded). Rotating a bitangent (red) to the next inner convex hull edge produces a line pq (blue) that can be used to represent the line partition as an anchored line partition.

the crossing. If one of the bitangents touches more than one vertex of A, we choose p and r to be one of these vertices that makes the inner chain as short as possible, as we have done in the figure by choosing p as shown rather than at the other vertex touched by its bitangent. Let q be the neighbor of p on the inner chain. Then we can rotate the bitangent that touches p until it contains segment pq; this rotation will swing the bitangent away from B, and it will not cross A because (at the end of the rotation) it coincides with a convex hull edge. So p and q are the pair of points we have been looking for. The line through convex hull edge pq (the blue line of the figure) separates the rest of A from B, and perturbing this line produces an anchored line partition (defined by the pair (p, q)) that matches the line partition of L. □

Corollary 3.13

For every finite configuration, the line partitions of the configuration are an invariant of the configuration. That is, every two sets of points with the same order type have the same line partitions.

Proof. For collinear configurations this is clear, and for noncollinear configurations, this follows from Lemma 3.12. □

The line partitions characterized here are closely related to a famous problem in discrete geometry, the *k-set problem*. We will not treat this problem in detail, because it does not involve a monotone parameter, but it can be formulated using line partitions as follows. We define a *halving partition* to be a line partition with equal numbers of points on both sides of the partition.

Open Problem 3.14

What is the maximum number of halving partitions that an n-point set (with n even) can have?

The best upper bound known for the problem is $O(n^{4/3})$,[9] while the best lower bound is only slightly superlinear.[10]

[9] Dey (1998). This bound is expressed in O-notation, which we will also use extensively for time bounds of algorithms; see Section 6.4 for details.

[10] Tóth (2001).

4 Subconfigurations

How many sides can a convex polygon with vertices in a 5×5 grid have? To avoid spoiling this puzzle we defer the answer to Figure 11.3. Just as one set of points may be a subset of another, one configuration may be a subconfiguration of another. The puzzle seeks a POLYGON subconfiguration of GRID$(5, 5)$. We define subconfigurations more carefully in the rest of this chapter. In Chapter 5, we will use them to define monotone properties and parameters and to find obstacles, the subconfigurations that prevent a given configuration from having a property.

4.1 Definitions

Definition 4.1

A *subconfiguration* of a configuration S is any configuration that can be obtained from a realization of S by removing zero or more points.[1] If T is a subconfiguration of S, an *instance* of T in S is a one-to-one matching of points in T to a subset of points in S so the matched points have the same orientations in T and in S. Each configuration is a subconfiguration of itself; T is a *proper* subconfiguration of S if it is a subconfiguration but unequal to S.

[1] For a different but related ordering on configurations, connecting configurations of the same size rather than different sizes, see Pilz and Welzl (2015).

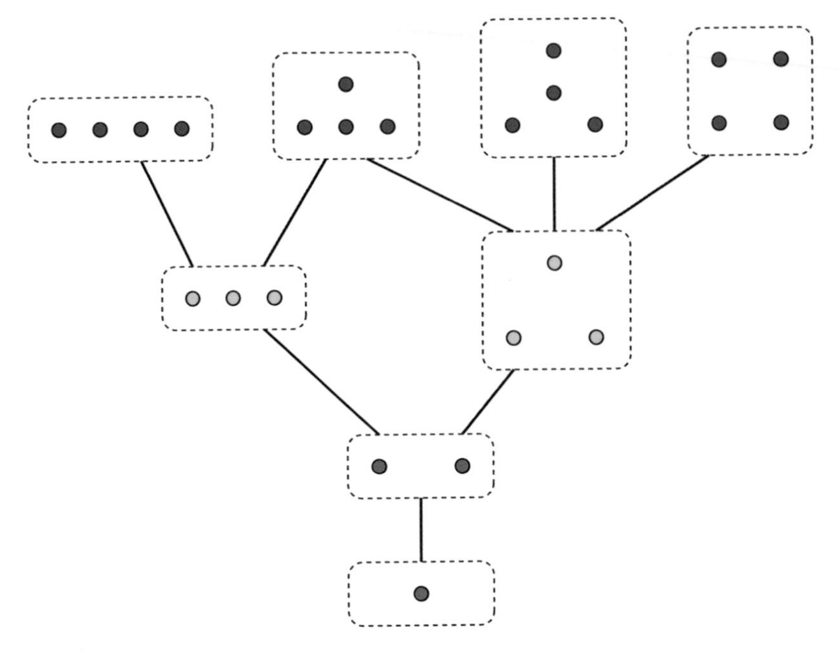

Figure 4.1. Subconfiguration relations on the configurations of up to four points

Example 4.2

Figure 4.1 shows all of the configurations on one to four points, with lines indicating pairs of subconfigurations that differ from each other by the addition or removal of a single point. In this diagram, one configuration is a subconfiguration of another if the two configurations are connected by a monotone path. For instance, the one-point configuration at the bottom of the diagram is a subconfiguration of all of the other configurations.

Note that the subconfigurations of a configuration T are *not* subsets of T, because T is not a set of points (it is an equivalence class of sets of points).

4.2 Partial Orders and Quasi-Orders

The subconfiguration relation between configurations forms an example of a *partial order*. This means that, if we write $S \hookrightarrow T$ when S is a subconfiguration of T,[2] then this relation $\hookrightarrow$ obeys the following three axioms:

[2] This $\hookrightarrow$ notation, both for point sets and configurations, is from Károlyi and Tóth (2012).

- Reflexivity: For all configurations X, $X \hookrightarrow X$.
- Anti-symmetry: For all configurations X and Y, if $X \hookrightarrow Y$ and $Y \hookrightarrow X$, then $X = Y$.
- Transitivity: For all configurations X, Y, and Z, if $X \hookrightarrow Y$ and $Y \hookrightarrow Z$, then $X \hookrightarrow Z$.

Other familiar partial orders include the subset relation $\subseteq$ on sets and the comparison relation $\leq$ on numbers. If we omit the requirement of anti-symmetry, we get a more general class of orderings called the *quasi-orders*.

Example 4.3

The relation on finite sets of points, defined by setting $A \hookrightarrow B$ whenever the configuration of A is a subconfiguration of the configuration of B, is a quasi-order but not a partial order.

4.3 Antichains

We will be studying the *obstacles* of properties of configurations. These are subconfigurations that prevent a configuration from having a property; we define them more carefully in Section 5.2. The obstacles for any given property will always form an *antichain*, a set of configurations none of which is a subconfiguration of any other. For, if one obstacle were a subconfiguration of another, it would be redundant: we could remove it from the collection of obstacles without changing the property defined by those obstacles.

Example 4.4

For any integer s, the configurations of size equal to s form an antichain. They are the obstacles for the property of having fewer than s points.

The antichains of Example 4.4 all have a finite number of configurations in them, but this is not true of some other antichains.

Example 4.5

The set of all sawtooth configurations,

$$\{\textsc{sawtooth}(n) \mid n \geq 3\},$$

forms an infinite antichain. Removing any points from one of the sawtooth configurations results in a configuration where at least one convex hull edge is lacking an interior point near its midpoint. This is true for each new convex hull edge formed by the removal and for each existing edge whose near-midpoint was removed. Therefore, when m and n are two different integers, it is not possible for $\textsc{sawtooth}(n)$ to be a subconfiguration of $\textsc{sawtooth}(m)$. That is, no sawtooth configuration is a subconfiguration of another one.

4.4 Well-Quasi-Ordering

One way to approach the question of which properties can be characterized by finitely many obstacles is to use the theory of *well-quasi-orderings*.[3]

Definition 4.6

A *well-quasi-ordering* is a quasi-order defined by a binary relation $\sqsubseteq$ such that, for every infinite sequence $x_1, x_2, \ldots$ of its elements, there exist indexes i and j for which $i < j$ and $x_i \sqsubseteq x_j$.

A well-quasi-ordering that is also a partial order may be called a *well-partial-ordering*, but we will not usually distinguish between well-quasi-orderings and well-partial-orderings. A quasi-order can fail to be a well-quasi-ordering in either of two ways: it can contain an infinite descending chain (a strictly descending sequence of elements) or an infinite antichain.

The partial order of configurations and subconfigurations acts like a well-quasi-ordering in one way: it has no infinite descending chains. This is because every descending sequence removes at least one point at each step, and each configuration has finitely many points to remove. However, as Example 4.5 shows, there are infinite antichains. Therefore, the partial order of configurations and subconfigurations is not a well-quasi-ordering. But as we will see, some natural subclasses of configurations are well-quasi-ordered. This implies that monotone properties within these classes have a finite number of obstacles.

In proving the well-quasi-ordering of certain classes of configurations, it will be helpful to use the following two standard lemmas. The first concerns k-tuples of nonnegative numbers, ordered by domination. In this ordering, for any two tuples of numbers v and w, we have $v \leq w$ when each coefficient of v is less than or equal to the corresponding coefficient of w.

[3] Kruskal (1972).

Lemma 4.7 (Dickson's lemma (Dickson, 1913))

For every integer k, the set of k-tuples of nonnegative integers is well-quasi-ordered by domination.

The next is a special case of Higman's lemma (Higman, 1952) on sequences of well-ordered values. A *subsequence* of a sequence is formed by taking the values from any subset of its positions in the same order. The positions chosen for the subsequence need not be consecutive.

Lemma 4.8

For every finite alphabet Σ, the set Σ^* of sequences of members of Σ, ordered by subsequences, is a well-quasi-order.

If we refine any well-quasi-ordering by adding more order relations between previously incomparable pairs of elements, producing another quasi-order, then the result must again be a well-quasi-ordering. For, if every infinite sequence of the original well-quasi-ordering has an ascending pair, then the same is true for the refined ordering. For instance, from the special case of Higman's lemma above, one can infer that the cyclic sequences of members of a finite alphabet are also well-quasi-ordered, because a cyclic sequence can be viewed as an equivalence class of linear sequences (the ones obtained by breaking the cycle at different points) and the subsequence quasi-order on the members of these equivalence classes is a refinement of the subsequence ordering on sequences.

5 Properties, Parameters, and Obstacles

With the notions of configurations and subconfigurations in hand, we are now ready to define monotone properties and monotone parameters.

5.1 Definitions

Definition 5.1

We define a *property* of configurations to be something that is true of some configurations and false for others, but whose truth or falsity depends only on the configuration and not on the set of points that realizes it. A property is *monotone* if, whenever a configuration has the property, so do all its subconfigurations.

We have already discussed the monotonicity of the happy ending problem.

Example 5.2

Containing the vertices of a convex quadrilateral is not a monotone property, but *not* containing a convex quadrilateral is monotone.

We may formalize a property either as a subset of the set of all configurations or as the *characteristic function* of this subset, a function of configurations that is true for the configurations with the property and false for the ones without it.

Definition 5.3

A *parameter* of configurations is a function from configurations to nonnegative integers. It is *monotone* if removing points from a configuration cannot cause the function value to increase.

Example 5.4

The size $|S|$ of a configuration S is a monotone parameter, because all realizations of S have the same number of points, and because taking away points reduces the size.

We can also form properties from parameters.

Example 5.5

Whenever ρ is a monotone parameter of configurations, and k is a positive integer, then the property of a configuration S of having $\rho(S) < k$ is a monotone property of configurations. For instance, the property of having size at most four is monotone.

Many classical problems in the discrete geometry of point sets can be formulated in terms of monotone properties and monotone parameters of configurations.

5.2 Obstacles

Every monotone property of configurations can be characterized by its *obstacles*, the smallest configurations that do not have the property.

Definition 5.6

If P is a monotone property of configurations, then the *obstacles* of P are the configurations that do not have P, but for which all proper subconfigurations do have P. P has *bounded obstacles* if there are finitely many obstacles for P.

Example 5.7

The property of being in general position, discussed at the end of Section 3.2, is monotone: every subconfiguration of a general-position configuration remains in general position. This property is violated when a set of points or configuration contains three points on a single line. Thus, the configuration of three points on a line is the only obstacle for being in general position. Because it has only one obstacle, the property of being in general position has bounded obstacles.

We can also use obstacles to define monotone properties.

Definition 5.8

If $C_1, C_2, \ldots$ are configurations, we define

$$\textsc{forbidden}(C_1, C_2, \ldots)$$

to be the property of not having any of $C_1, C_2, \ldots$ as subconfigurations. We write

$$\textsc{forbidden}(C_1, C_2, \ldots; S)$$

for the value of this property on configuration S: false if S contains a subconfiguration C_i or true otherwise.

Then the obstacles for $\textsc{forbidden}(C_1, C_2, \ldots)$ are the minimal elements of $C_1, C_2, \ldots$. In particular, if $C_1, C_2, \ldots$ is any finite set of configurations, then $\textsc{forbidden}(C_1, C_2, \ldots)$ automatically has bounded obstacles.

Observation 5.9

There are uncountably many monotone properties of configurations.

Proof. Because there are infinitely many sawtooth configurations, there are uncountably many subsets of the set of sawtooth configurations. Each such set S defines a different monotone property $\textsc{forbidden}(S)$. □

We will study the sets of obstacles for many properties defined from bounded parameter values (as in Example 5.5). In that example, we considered any monotone parameter ρ and positive integer k, and from them defined a monotone property, the property of having $\rho(S) < k$. The number of obstacles for these properties can be large, and precisely describing these obstacles can be difficult. Therefore, it will often be more convenient to study the size of these obstacles rather than the obstacles themselves.

Definition 5.10

If ρ is a monotone parameter of configurations, then the *obstacle size* of ρ is the function that maps a parameter value k to the largest number of points in an obstacle for the property $\rho(S) < k$. If this property does not have bounded obstacles, we write ∞ for the obstacle size. A monotone parameter has *finite obstacle size* if its obstacle size is always a finite number, and *polynomial obstacle size* if in addition this number is at most a polynomial function of k.

Example 5.11

The parameter $|S|$ that measures the number of points in a configuration S has polynomial obstacle size. For, the obstacles for $|S| < k$ are all the k-point configurations, and these have size k.

For an example of a parameter with bounded obstacle size but not polynomial obstacle size, see Theorem 12.9.

5.3 Parameters from Obstacles

Just as we defined the property FORBIDDEN$(C_1, C_2, \ldots)$ from an arbitrary list of obstacles $C_1, C_2, \ldots$, we can also define parameters from explicit lists of obstacles.

Definition 5.12

If $C_1, C_2, \ldots$ are configurations, we define the following parameters:

- Parameter AVOIDS$(C_1, C_2, \ldots)$ maps a configuration S to the largest size of a subconfiguration of S with property FORBIDDEN$(C_1, C_2, \ldots)$. That is, it is the size of the largest subconfiguration that avoids having any of $C_1, C_2, \ldots$ as sub-subconfigurations.

- Parameter HITTING$(C_1, C_2, \ldots)$ is the smallest size of a subconfiguration H of S that hits (has a nonempty intersection with) every instance of $C_1, C_2, \ldots$ as a subconfiguration of S.
- Parameter PARTITION$(C_1, C_2, \ldots)$ is the minimum number of subconfigurations that S must be partitioned into so that each of them has property FORBIDDEN$(C_1, C_2, \ldots)$.

As with FORBIDDEN$(C_1, C_2, \ldots; S)$, we write

$$\text{AVOIDS}(C_1, C_2, \ldots; S),$$

$$\text{HITTING}(C_1, C_2, \ldots; S), \text{ and}$$

$$\text{PARTITION}(C_1, C_2, \ldots; S)$$

for the values of these parameters on a specific configuration S.

Example 5.13

The parameter $|S|$ that measures the number of points in a configuration S can equivalently be defined in each of the following ways:

- $|S| = \text{AVOIDS}(\emptyset; S)$. The size of any configuration S equals the largest size of a subconfiguration of S that avoids the empty set of obstacles.
- $|S| = \text{HITTING}(C_1; S)$ where C_1 is the unique one-point configuration. For, the size of any configuration S is the number of points in the smallest subconfiguration of S that has a nonempty intersection with each copy of the one-point configuration.
- $|S| = \text{PARTITION}(C_2; S)$ where C_2 is the unique two-point configuration. That is, the size of S equals the minimum number of subconfigurations into which we can partition S so that each subconfiguration has at most one point.

5.4 Relation between Parameters

There is a simple relation between the two types of parameters AVOIDS and HITTING, defined above. For, HITTING$(C_1, C_2, \ldots)$ is the smallest number of points that need to be removed from S in order to create a subconfiguration with property FORBIDDEN$(C_1, C_2, \ldots)$, while AVOIDS$(C_1, C_2, \ldots)$ is just the size of that remaining subconfiguration. Therefore, we have the following observation.

Observation 5.14

For every configuration S and every set of obstacles $C_1, C_2, \ldots,$

$$\text{HITTING}(C_1, C_2, \ldots; S) = |S| - \text{AVOIDS}(C_1, C_2, \ldots; S).$$

We can also relate these two parameters with the third type of parameter, PARTITION, by inequalities rather than equality.

Observation 5.15

Let $C_1, C_2, \ldots$ be any set of obstacles all of which have at least k points, for some integer $k > 1$. Then, for every configuration S, we have the inequalities

$$\left\lceil \frac{|S|}{\text{AVOIDS}(C_1, C_2, \ldots; S)} \right\rceil \leq \text{PARTITION}(C_1, C_2, \ldots; S) \leq 1 + \left\lceil \frac{\text{HITTING}(S)}{k-1} \right\rceil .$$

Proof. The first inequality follows from the fact that any partition into subconfigurations of size at most a must use at least n/a subconfigurations. The second inequality follows from the fact that the points of S can be partitioned into a hitting set of minimal size and an avoiding set containing all the remaining points (Observation 5.14), and that any partition of the hitting set into $(k-1)$-point subsets avoids all the given obstacles. □

5.5 Comparing Parameters

We would like to be able to compare different parameters with each other, to answer the following questions. Which parameter has more descriptive power? If we restrict our attention to configurations with bounded parameter values, do we get more configurations in our restricted set with one parameter or with the other? To do this, it will be convenient to compare parameters not by their actual values, but by whether we can bound one parameter by a function of the other.[1]

[1] For a related but different notion of comparison of parameters, based on linear inequalities rather than arbitrary functional inequalities, see Komusiewicz and Niedermeier (2012).

Definition 5.16

If two parameters $\textsc{parameter}_1$ and $\textsc{parameter}_2$ are each upper-bounded by a function of the other parameter, we write this relation as

$$\textsc{parameter}_1 \doteq \textsc{parameter}_2.$$

If $\textsc{parameter}_1$ is upper-bounded by a function of $\textsc{parameter}_2$, but not vice-versa (that is, there exist families of instances for which $\textsc{parameter}_1$ is bounded but $\textsc{parameter}_2$ is unbounded), then we write this relation as

$$\textsc{parameter}_1 \ll \textsc{parameter}_2.$$

Example 5.17

The size of a configuration, and the number of distinct pairs of points in the configuration, are equivalent parameters according to the $\doteq$ relation. For, if we let n denote the size and p denote the number of pairs in any given configuration, then n can be upper-bounded by a function of p as

$$n \leq 1 + \sqrt{2p}$$

and conversely p can be upper-bounded by a function of n as

$$p \leq \frac{n(n-1)}{2}.$$

This example shows that equivalent parameters do not always have the same growth rate as each other.

Observation 5.18

The relation $\doteq$ is an equivalence relation, the relation $\ll$ is transitive and irreflexive, and parameters that are equivalent according to $\doteq$ have the same relations as each other with respect to $\ll$.

That is, the relation $\leq$ on parameters that is true when

$$\textsc{parameter}_1 \ll \textsc{parameter}_2$$

or

$$\textsc{parameter}_1 \doteq \textsc{parameter}_2,$$

and false otherwise, is a quasi-order. We will not investigate the structural properties of this quasi-order (such as whether parts of it form a well-quasi-order), but we will investigate the relationships between specific pairs of parameters.

For a general result describing which parameters $\textsc{avoids}(C_1, C_2, \dots)$ are equivalent to the size parameter, and which are less than the size, see Theorem 11.20.

We conclude this section with some observations on what happens when we combine two parameters, for instance by taking their sum. The exact choice of which operation to use to combine them will often not matter very much.

Observation 5.19

Let $\textsc{parameter}_1$ *and* $\textsc{parameter}_2$ *be any two nonnegative integer parameters. Then we have*

$$\max(\textsc{parameter}_1, \textsc{parameter}_2) \doteq \textsc{parameter}_1 + \textsc{parameter}_2.$$

If the parameters are both nonzero, then these two combinations are also equivalent to the product

$$\textsc{parameter}_1 \cdot \textsc{parameter}_2.$$

Proof. These equivalences follow from the inequalities

$$\begin{aligned}\max(&\textsc{parameter}_1, \textsc{parameter}_2)\\ &\le \textsc{parameter}_1 + \textsc{parameter}_2\\ &\le 2\max(\textsc{parameter}_1, \textsc{parameter}_2)\end{aligned}$$

and, when both are nonzero,

$$\begin{aligned}\max(&\textsc{parameter}_1, \textsc{parameter}_2)\\ &\le \textsc{parameter}_1 \cdot \textsc{parameter}_2\\ &\le \max(\textsc{parameter}_1, \textsc{parameter}_2)^2.\end{aligned}$$

□

Observation 5.20

If $\textsc{parameter}_1 \doteq \textsc{parameter}_2$, *then*

$$\min(\textsc{parameter}_1, \textsc{parameter}_2) \doteq \max(\textsc{parameter}_1, \textsc{parameter}_2).$$

Otherwise,

$$\min(\textsc{parameter}_1, \textsc{parameter}_2) \ll \max(\textsc{parameter}_1, \textsc{parameter}_2).$$

Proof. In both cases,

$$\min(\text{PARAMETER}_1, \text{PARAMETER}_2) \leq \max(\text{PARAMETER}_1, \text{PARAMETER}_2).$$

The instances in which the min is bounded and the max is unbounded are exactly those for which one of the two parameters is bounded and the other is unbounded, so such a family of instances exists if and only if the two parameters are not equivalent. □

6 Computing with Configurations

How many triangles are there in the configuration of Figure 6.1? To answer this, one needs either some mathematical insight or tedious calculation. But performing tedious calculations is something computers do well. So, how could a computer be used to solve this puzzle?

6.1 Input Representation

Before we ask how to design a program to find triangles or other subconfigurations in larger configurations, we need to consider how a configuration can be described as the input to a program. In the case of Figure 6.1, the configuration is already shown with grid lines, making it easy to represent its points using integer Cartesian coordinates. But not every configuration can be represented in this way (see Chapter 13 for details). Even the configurations that do have integer coordinate representations sometimes need huge numbers for those coordinates, forcing algorithms with this kind of input to use multiprecision arithmetic and complicating their analysis.[1]

Instead, for algorithms on configurations, we will generally assume only that our algorithms have some way to identify the points of a configuration and to determine the orientations of triples of points. Algorithms that access their input in this limited way are more versatile. They can handle inputs using explicit coordinates for the points, as long as we can compute orientations from coordinates. But they can also handle a three-dimensional matrix of orientations as input. It would even be possible for such an algorithm to take as its

[1] Goodman et al. (1989).

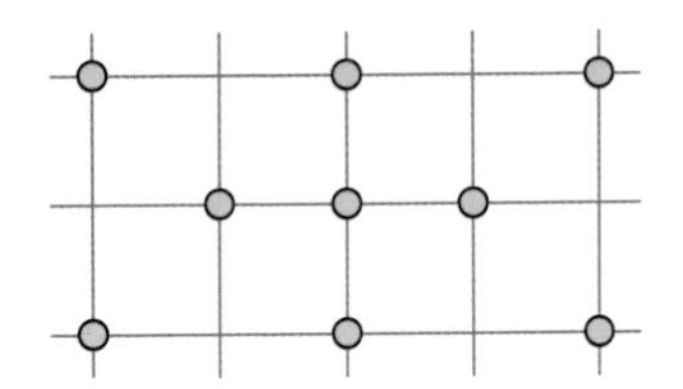

Figure 6.1. How many triangles are there in this configuration?

input a number of points and a pointer to a subroutine that maps triples of indexes to the orientation of the points with those indexes. When analyzing these algorithms, we will assume that each orientation test takes constant time, but if not, the time bounds we state should be multiplied by the time per orientation test.

Despite being restricted in this way, it is still possible for an algorithm to carry out many important geometric tasks, such as the construction of the convex hull of the input points. Recall from Chapter 1 that the convex hull is a convex polygon, the smallest convex polygon that contains the set. Figure 6.2 shows an example. For points that are not all on a single line, the convex hull may be determined from the order type, as follows.

- A line through two points p and q supports an edge of the convex hull if and only if all triples (p, q, r) have nonnegative orientation or all have nonpositive orientation.
- A point of the set is a vertex of the convex hull if and only if it belongs to two distinct supporting lines.
- Each edge of the convex hull is the line segment that lies on a supporting line and has the two vertices of that line as its endpoints.

By applying reasoning of this type one can design an algorithm that computes the convex hull in near-linear time, for an input of n points, and that accesses its input set of points purely by means of orientation queries.[2]

6.2 Computing the Order Type

Continuing our quest for a computer program to find triangles, suppose that our hypothetical program has found a triple of points that we think might be a triangle. How could we tell for sure? The answer is yes when the orientation of the triple is nonzero, but how can we compute the orientation of three points from their Cartesian coordinates? A subroutine for doing this will be useful

[2] Knuth (1992). See especially pp. 47–51. For additional research on algorithms using abstract order types as input, see Aichholzer et al. (2013).

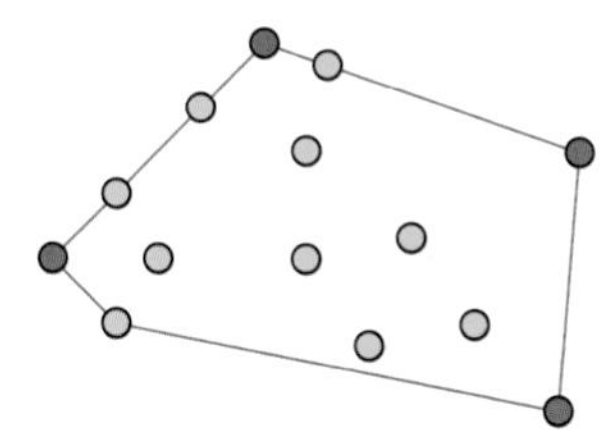

Figure 6.2. The convex hull of a set of points (the convex polygon shown by the blue points and line segments) may be determined purely from its order type.

more generally, as the "glue" needed to adapt a coordinate-based input representation to an algorithm that uses orientation tests as its primitive operations. Such a subroutine will also, in effect, compute the order type of a set of points, because an order type is nothing but a function that maps triples of points to their orientations.

Orientations can be computed by a simple formula that uses the sign of a matrix determinant. Any three points with the coordinates (x_i, y_i) (for $i = 1, 2,$ or 3), have the orientation

$$\text{orientation} = -\operatorname{sign}\det\begin{bmatrix} x_1 & y_1 & 1 \\ x_2 & y_2 & 1 \\ x_3 & y_3 & 1 \end{bmatrix},$$

where we define sign x to be -1, 0, or $+1$, respectively, as $x < 0$, $x = 0$, or $x > 0$. Expanding the determinant in this formula into a polynomial gives the equivalent formula

$$\text{orientation} = -\operatorname{sign}(x_1y_2 - x_1y_3 + x_2y_3 - x_2y_1 + x_3y_1 - x_3y_2).$$

So a subroutine for computing orientations needs only to compute this polynomial, determine its sign, and return the result.

An explanation for this otherwise-mysterious matrix formula can be found in the following ideas. If we fix the positions of two points (x_1, y_1) and (x_2, y_2), but let the third point (x_1, y_3) remain variable, then the determinant becomes a linear function of the remaining variables x_3 and y_3. This linear function vanishes when the third row of the matrix is a linear combination of the other two rows, which is true when point (x_3, y_3) belongs to the line through the other two points. As a linear function, it must have positive sign on one side of this line and negative sign on the other side of the line. If we check simple examples such as the triangle with vertex coordinates $(0, 0)$, $(0, 1)$, and $(1, 0)$, we find that the sign of the determinant needs to be negated to produce the correct result.

The determinant itself is also meaningful. Its absolute value is twice the area of the triangle determined by the three points. When the three points are on a line, the triangle is degenerate and its area is zero.

6.3 Finding Subconfigurations

Suppose that we want to find the instances of a fixed configuration T as a subconfiguration of other configurations.[3] For instance, in the puzzle that started this chapter, T is a triangle. How hard is this problem? The answer is: not very hard, in theory. It can be solved in *polynomial time.* This means that there is an algorithm that solves the problem whose number of steps is at most a polynomial function of the input size.

Algorithm 6.1 (Finding matching subconfigurations)

Given a configuration T of size k, and another configuration S of size $n \geq k$, perform the following steps:

1. Number the points of T from 1 to k.
2. For each ordered k-tuple K of distinct points from S, perform the following steps:
   a. Loop through all of the triples of indexes (a, b, c) with $1 \leq a < b < c \leq k$. For each triple:
      i. Compute the orientation of the points with indexes a, b, and c in T.
      ii. Compute the orientation of the points with indexes a, b, and c in K.
      iii. If the two orientations differ, break out of the inner loop and go on to the next k-tuple.
   b. If we did not break out of the loop, output K as one of the instances of T in S.

Observation 6.2

For any fixed configuration T, Algorithm 6.1 finds all instances of T as a subconfiguration of an input configuration S in polynomial time.

Proof. The outer loop of the algorithm considers fewer than n^k k-tuples of points, and the inner loop considers $\binom{k}{3}$ triples of indexes. Each iteration of the inner loop involves a constant number of orientation tests and a constant

[3] There is a substantial body of related work on testing whether a point set can be transformed to become a subset of a larger point set. See, e.g., Huttenlocher and Kedem (1990); Sprinzak and Werman (1994); Chew et al. (1997).

amount of additional work comparing the orientations. Therefore, the total time for the algorithm is proportional to $n^k\binom{k}{3}$.

When T is fixed, k is also fixed, and $n^k\binom{k}{3}$ is a polynomial in n. For instance, for $k = 4$ it is the polynomial $4n^4$. (It would not be a polynomial when k is treated as a variable and this time bound is considered as a function of k, however, because k appears in the exponent of n.) Therefore, for fixed T, the time for this algorithm is polynomial. □

It follows that any property with bounded obstacles can automatically be recognized in polynomial time, by using Observation 6.2 to test whether any of the obstacles is present as a subconfiguration of an input configuration.

If we use this method to count triangles in Figure 6.1, it will give the answer 222 instead of what we might expect, 74. Each triangle in Figure 6.1 forms six ordered 3-tuples K in Algorithm 6.1, three of which match the orientation of POLYGON(3) and the other three of which have the opposite orientation. Therefore, although Figure 6.1 has 74 triangles, each of them counts as three different instances of POLYGON(3). The cubic time for using Algorithm 6.1 to count triangles can be improved; see Corollary 8.7.

6.4 *O*-Notation

It is standard in the analysis of algorithms to use a mathematical notation called O-notation to simplify the time bounds of algorithms. Rather than writing sentences like "the total time for the algorithm is proportional to $n^k\binom{k}{3}$," as in the proof of Observation 6.2, we will write more simply that "the algorithm takes time $O(n^k)$."

Formally, for two functions f and g, we write[4]

$$f = O(g)$$

to mean that there exists a constant $C > 0$ such that, for all sufficiently large n, $f(n) \leq Cg(n)$.

In practice, O-notation is mainly used to simplify expressions. When a function is the sum of two terms, one of which grows more quickly than the others, we keep only the more quickly growing term, and when a function is the product of two factors, one of which is bounded, we keep only the unbounded factor. So we can simplify $5n^3 + 3n^2 + 6n - 42$ by keeping only its leading term $5n^3$ and stripping off the constant factor 5, giving $5n^3 + 3n^2 + 6n - 42 = O(n^3)$. The ability of O-notation to eliminate constant factors is especially useful in the analysis of algorithms, because otherwise the speed of the

[4] We follow the computer science conventions for O-notation and its relatives, which differ somewhat from the conventions of mathematical analysis where this notation arose. To make sense mathematically, "$= O$" should be interpreted as a single piece of notation for a relation between two functions, rather than an equality sign followed by an operator on one function.

particular machine on which we run the algorithm would be a factor in its runtime, making it difficult to perform analysis in a machine-independent way.

We will occasionally use some related notation involving the letters o, ω (the Greek letter omega), Ω (capital omega), and Θ (capital Greek letter theta). We summarize their meanings in the following table. In all cases, the definition includes an implicit "for all sufficiently large n," where "sufficiently large" may depend on the choice of the constant C.

Notation	Mathematical meaning	Intuitive meaning
$f = O(g)$	$\exists C > 0 : f(n) \leq Cg(n)$	f is at most proportional to g
$f = o(g)$	$\forall C > 0 : f(n) \leq Cg(n)$	f grows less quickly than g
$f = \Omega(g)$	$\exists C > 0 : f(n) \geq Cg(n)$	f is at least proportional to g
$f = \omega(g)$	$\forall C > 0 : f(n) \geq Cg(n)$	f grows more quickly than g
$f = \Theta(g)$	$f = O(g)$ and $f = \Omega(g)$	f and g grow at the same rate

6.5 Property Testing

It is not always necessary to look at the whole input to an algorithm in order to compute something useful about it. *Property testing algorithms* aim to distinguish inputs that have some property from inputs that are far from having the property, very quickly, by looking at only parts of an input rather than the whole input. Often these methods are based on random sampling, and they may occasionally select an unlucky sample that causes them to make a mistake. The goal is to develop algorithms that keep both the sample size and the rate of mistakes low.[5] To test monotone properties of configurations, we will look at a special kind of property testing algorithm whose steps are fixed, with the only design choice being how large of a sample to choose.

Algorithm 6.3 (Sample-based property testing)

We define a *sample-based property testing algorithm* for a monotone property Π of configurations to be an algorithm that, on an input configuration S, performs the following steps:

1. Compute a sample size k as a function of the size $|S|$ of the input.
2. Choose a sample configuration K of k points from S, uniformly at random among all subsets of k points.
3. Determine whether K has property Π.[6]
4. Return the result of the previous step as the output of the algorithm.

[5] Goldreich (2010).

[6] This step needs to be filled in differently for different properties, using an exact algorithm rather than a property testing algorithm.

Thus, these algorithms themselves are very simple. However, their analysis may be complex. To describe the accuracy of such an algorithm, we need a more formal definition of what it means to be far from having property Π.

Definition 6.4

For a given configuration S, property Π, and number ϵ in the range $0 \leq \epsilon \leq 1$, we say that S is ϵ-near to property Π if one can remove $\epsilon|S|$ points from S and obtain a subconfiguration with property Π. If not, we say that S is ϵ-far from property Π.

When the obstacles of Π are $C_1, C_2, \ldots$, a configuration S is ϵ-near to Π when AVOIDS$(C_1, C_2, \ldots; S)$ is large (close to the size of S) and HITTING$(C_1, C_2, \ldots; S)$ is small. Conversely, S is ϵ-far from Π when AVOIDS is small and HITTING is large.

Definition 6.5

There are two different type of errors that a property testing algorithm might make.

- A *false negative* occurs when an input S has property Π, but a property testing algorithm reports that it does not have the property.
- A *false positive* (for a given distance ϵ with $0 < \epsilon < 1$) occurs when an input S is ϵ-far from having property Π, but a property testing algorithm reports that it does have the property.

We wish to keep the rates of both of these kinds of errors small. However, for inputs that are ϵ-near to property Π but do not have the property, the algorithm will be allowed to report either possible answer. We do not consider this to be an error, because these inputs are not the ones that the property testing algorithm is designed to handle.

Observation 6.6

A sample-based property testing algorithm for a monotone property can never produce any false negatives.

Proof. The property Π that we are testing is assumed to be monotone. Therefore, if S has the property, so does the subconfiguration K tested by the algorithm. □

Definition 6.7

For a number p in the range $0 \leq p \leq 1$, we say that a property testing algorithm for property Π and distance ϵ achieves false positive rate p if, for every input S that is ϵ-far from Π, the algorithm has probability at most p of producing a positive answer.

We will generally aim for the false positive rate to be a constant between 0 and 1, independent of the input size. If we can achieve this, for a given sample size, we will also be able to achieve a false positive rate that converges to zero in the limit as the size of the input increases, by making a corresponding increase in the sample size.

With these definitions in hand, our goal becomes one of determining, for given numbers ϵ and p, the smallest sample size k for which a sample-based property testing algorithm can achieve false positive rate p. Sample size $k = |S|$ will work: if our samples equal the whole input, there can be no false positives. However, we would like to use a sample size substantially smaller than $|S|$. If we can do so, the sample-based property testing algorithm will automatically be faster than an algorithm that determines exactly whether the whole configuration has the given property, because it is performing a similar computation on a smaller subconfiguration.

Theorem 6.8

Let $C_1, C_2, \ldots$ be a finite set of obstacles, whose maximum size is s, and let ϵ and p be numbers in the range $0 < \epsilon < 1$ and $0 < p < 1$ (both independent of the input size). Then there is a sample-based property testing algorithm for the property $\Pi = \text{FORBIDDEN}(C_1, C_2, \ldots)$ whose sample size, on configurations of size n, is $O(n^{1-1/s})$ and whose false positive rate for configurations ϵ-far from this property is at most p.

Proof. The analysis of sample-based property testing is complicated by the fact that points of S are not sampled independently: choosing one point makes choosing any other one less likely. To simplify the analysis, we relate this algorithm to a modified algorithm in which samples are independent. The algorithm performs the following steps.

Algorithm 6.9 (Modified sample-based property testing)

1. For a constant c to be determined later, select a random sample R of the points of the input configuration S by including each point independently with probability $c/n^{1/s}$.
2. Compute a sample size $k = O(n^{1-1/s})$ such that, with probability at least $1 - p/2$, $|R| < k$.
3. Compare $|R|$ with k. If $|R| > k$, return the result that S has property Π (regardless of whether it actually does).
4. Otherwise, test whether R has property Π and return the result.

Let us compare the behavior of this algorithm to the behavior of a sample-based property testing algorithm that uses the same sample size k, on an input S that is ϵ-far from Π. To be more specific, let R be the sample of the modified algorithm, and suppose that (when $|R| \leq k$) the sample-based algorithm constructs its sample by adding $k - |R|$ additional randomly chosen points to R. This assumption does not affect the uniform randomness of the sample. Then, with probability $p/2$, the modified algorithm generates a sample that is too big, and produces a false positive. In the same cases, the sample-based algorithm might or might not produce a false positive. And in the remaining cases, any sample that would cause the sample-based algorithm to produce a false positive is a superset of a sample that would cause the modified algorithm to also produce a false positive. In every case, whenever the sample-based algorithm produces a false positive, so does the modified algorithm (but not necessarily vice versa). This means that the false positive rate of the sample-based algorithm is at least as good as the false positive rate of the modified algorithm. As long as some value of c causes the modified algorithm to achieve false positive rate p, the same will be true of the sample-based algorithm.

To determine what value of c we should use, consider an arbitrary configuration S of size n that is ϵ-far from property Π. Let X be a maximal family of disjoint instances of the obstacles $C_1, C_2, \ldots$ in S. Here, "disjoint" means that no two of these instances have a point in common, and "maximal" means that we cannot find any other instances that are not in X and are disjoint with the instances in X. Then, removing all points in X from S would leave a subconfiguration without any instances of the obstacles: the obstacles that overlap X each have at least one point removed, and no other obstacles exist. Therefore, this subconfiguration would have property Π. It follows that the total number of points in X is at least ϵn, for otherwise S would not be ϵ-far from Π. Therefore, also, the number of instances in X is at least $\epsilon n/s$.

Let $X_1, X_2, \ldots$ be the instances of obstacles in X. Then for any single instance X_i, the probability that all points in X_i are included in the random sample R is

$$\left(\frac{c}{n^{1/s}}\right)^{|X_i|} \geq \frac{c^s}{n},$$

because $|X_i| \leq s$. If all points of X_i are included, then R will not have property Π and the algorithm will not produce a false positive. And because the instances are disjoint and we included points in R independently, distinct instances in X have independent probabilities of being completely included in R. Therefore, the rate of false positives that happen when R has property Π is at most

$$\left(1 - \frac{c^s}{n}\right)^{\epsilon n/s} < \exp \frac{-c^s \epsilon}{s}.$$

If we choose

$$c = \left(\frac{s}{\epsilon} \ln \frac{2}{p}\right)^{1/s} = O(1),$$

this type of false positive will happen with probability at most $p/2$, and the total false positive rate of the modified property testing algorithm will be at most p. Therefore, by the above reasoning, the sample-based property testing algorithm for this sample size will also achieve false positive rate p. □

As this theorem shows, we can always test monotone properties with finitely many obstacles using a sublinear sample size. Later, in Theorem 8.12, we will see that in some cases, the sample size can be significantly smaller, even constant. On the other hand, Theorem 9.9 and Corollary 11.9 provide examples of properties for which the polynomial dependence of sample size on input size in Theorem 6.8 is necessary.

Open Problem 6.10

Is there a natural monotone property Π of configurations and constants ϵ and p such that no sublinear sample-based property testing algorithm for Π can achieve a false positive rate p for configurations ϵ-far from Π?

If such a property exists, it would need to have infinitely many obstacles.

7 Complexity Theory

When we find ourselves unable to come up with an efficient algorithm for a computational problem, it is often possible to find an explanation (if not a rigorous proof that the algorithm does not exist) using the techniques of *computational complexity theory.*

7.1 Decision Problems and Complexity Classes

A computational problem can be described by a mathematical specification of its input and of the desired output for each input. This specification does not describe how to compute the output efficiently; that is the task of the algorithm designer. Within complexity theory, a special place is taken by the *decision problems,* the problems whose output is always either yes or no. A *complexity class* is a set of decision problems, usually grouped by how much time or memory their algorithms require.

In this theory, three of the most important complexity classes are P, NP, and the class of NP-complete problems. We describe them briefly here; for an accessible and more in-depth treatment, see Fortnow (2013).

- P is the class of problems with yes-or-no answers that can be solved by an algorithm whose runtime is polynomial in the number of bits of input. For example, one can test whether a list of points (given by Cartesian coordinates as binary numbers) contains any repeated points, by comparing each pair of points and checking whether they are equal. This algorithm takes time proportional to the square of its input length, a polynomial. It is also possible to solve the same problem more quickly, using sorting or hashing.

- NP is a larger class of yes-or-no problems, not all of which are believed to have efficient algorithms. However, whenever the answer to a problem in NP is "yes," there must exist a short and easily checked proof that the answer is yes. Many common puzzles, such as sudoku, can be formulated as problems in NP.[1] For such puzzles, the yes-or-no question is "does this puzzle have a solution?" or "if I add this step to my answer, will it still have a solution?" The proof that it does have a solution is the solution itself. Although the solution may be hard to find, it is easy to check that it is valid.
- The NP-complete problems are a special case of the problems in NP. They are defined by the property that instances of any other problem in NP can be translated (in polynomial time) into instances of the given NP-complete problem with the same answer. Therefore, if an NP-complete problem had an efficient algorithm, this translation could be used to provide an efficient algorithm for everything else in NP. The NP-hard problems have this same translation property, but might not be in NP themselves.
- There is a standard technique for proving that a problem is NP-complete, called a "reduction." We start with another problem already known to be NP-complete, and describe a translation from instances of the known-hard problem into instances of the given problem. Composing this translation with the translations from anything else in NP into the known-hard problem (which must exist if the other problem is NP-complete) shows that the target of the translation is also NP-hard. If it is also in NP, it is NP-complete.

Related to NP is another complexity class called $\exists\mathbb{R}$, which describes problems that can be represented (via a polynomial time translation) as the solution of a system of simultaneous equations in real-number arithmetic. Testing whether a three-dimensional matrix of values in $\{0, +1, -1\}$ forms the order type of a set of points is NP-hard and $\exists\mathbb{R}$-complete.[2]

We will also see another lesser-known complexity class called Σ_2^{P}.[3] Problems in NP can be solved by a technique called *brute force search* in which we loop through a collection of candidate proofs and check the validity of each candidate proof. If a valid proof is found, the result for the given instance is "yes," and otherwise it is "no." Each candidate proof has polynomial size and can be checked in polynomial time, but there may be exponentially many proofs to check. Σ_2^{P} can be described in the same way, but with a more complicated proof checker: its problems can be solved by a brute force search that loops through a collection of candidate proofs of polynomial size, but where determining that a proof is invalid is itself a problem in NP. Like NP, Σ_2^{P} has its complete problems, problems in Σ_2^{P} to which every other problem in Σ_2^{P} can be reduced. In Theorem 7.4 we will see a natural and fundamental Σ_2^{P}-complete problem on configurations and subconfigurations.

[1] Hearn and Demaine (2009). [2] Shor (1991).

[3] The notation for Σ_2^{P} indicates that it is at the second level of a system of complexity classes that has P and NP on its zeroth and first levels, respectively. For a survey of Σ_2^{P} and its complete problems, see Schaefer and Umans (2002).

However, for practical computation, the differences between NP, $\exists\mathbb{R}$, and Σ_2^P are subtle and not very important. Whenever a problem is complete for any of these classes, it means that it can be computed in exponential time and (although we do not know how to prove it) probably cannot be computed in polynomial time. For each of these classes, we prove completeness in the same way, by finding a polynomial-time translation from a known-complete problem to the problem we want to prove hard.

7.2 Approximation Algorithms

We will also consider the complexity of approximation of some of the problems we study. This means that we seek algorithms that run in polynomial time and that produce a solution that is valid but not necessarily optimal. The *approximation ratio* of one of these algorithms is a number ρ such that the algorithm's output is always within a factor of ρ of the exact answer. For instance, the number of colors needed to color a planar graph can be approximated accurately by always returning four as the answer. This is always valid – every planar graph can indeed be colored with four colors, as the famous four-color theorem[4] proves – but is not always optimal, as some planar graphs can be colored with fewer colors. Nevertheless, it is always within a factor of four of the correct number of colors, so its approximation ratio is at most four.[5]

When developing approximation algorithms, we seek an approximation ratio that is as close as possible to one, because that indicates that the algorithm's answer is as close as possible to optimal. In some cases, we can find a *polynomial-time approximation scheme*, an algorithm that can achieve an approximation ratio of $1 + \epsilon$ for any given number $\epsilon > 0$. However, not every problem can be approximated as accurately as that. Complexity theorists have defined the complexity class APX to capture this notion of whether an accurate approximation is possible.

APX is the class of NP optimization problems that have a bounded approximation ratio. That is, they have approximation algorithms that produce solutions that are always within a constant factor of optimal. A problem is APX-hard if every problem in APX can be translated into it by a reduction that preserves the existence of polynomial-time approximation schemes, and APX-complete if it is in APX and APX-hard. Being APX-complete implies that there is a constant c such that one can approximate the problem to within a factor of c, but no better. Many APX-complete problems are known.[6]

[4] Appel and Haken (1989); for surveys see Saaty and Kainen (1986) and Wilson (2014).
[5] By separately handling the graphs that can be colored by two or fewer colors, it is straightforward to improve this approximation ratio for planar graph coloring to 4/3.
[6] Alimonti and Kann (2000).

7.3 Parameterized Complexity

The field of *parameterized complexity* concerns problems defined not only by their intended input-output behavior but also by a *parameter* that describes how complicated the input is. For us, these parameters will typically be integer-valued functions of the inputs, with larger function values meaning that the input is more complicated. The algorithms that we study will not necessarily have access to the parameter values, but we will incorporate these values into their analysis. The goal of this area is to develop algorithms that can work well for inputs whose parameters are small, even when the input size is large. In this way, problems without a fully polynomial time algorithm can nevertheless find practical solutions. For recent reviews, see Downey and Fellows (2013) or Flum and Grohe (2006). We repeat the following standard definitions from this area.

Definition 7.1

A parameterized problem is *nonuniform fixed-parameter tractable* if there exists a constant c such that, for every constant value k of the parameter, the inputs of size n and parameter value k can be solved in time $O(n^c)$. Here *nonuniform* means that we can use different algorithms for different values of k. A problem is *fixed-parameter tractable* if, more strongly, there is a single algorithm that solves all inputs in time $O\big(n^c f(k)\big)$ for a computable function f.

A common way to show that a problem is fixed-parameter tractable is to use *kernelization*. This is an algorithmic technique in which we solve a problem in the following three-step process.

Algorithm 7.2 (General method of kernelization-based parameterization)

1. Transform the input to a smaller input, called a *kernel*.
2. Find the optimal solution to the kernel, for instance by a brute force search.
3. Transform the solution for the kernel into a solution to the original instance.

The steps that transform an input to a kernel and that transform the kernel's solution back into a solution to the original problem should both run in polynomial time, independent of the parameter. Additionally, the kernel should have a size bounded by a function of the parameter, independent of the original problem's size. If both of these things are true, the kernelization algorithm outlined above will be fixed-parameter tractable, because the time for the second step (the only step that is not polynomial) will automatically be bounded by a function of the parameter. Conversely, every problem that is fixed-parameter tractable can be solved using kernelization.[7] We will frequently use kernelization as a way of showing that certain computational problems on configurations are fixed-parameter tractable.

When we apply parameterized analysis to problems of computing the value of a monotone parameter of configurations, we will generally assume that the parameter of the analysis is the *natural parameter*, the one we are computing. If a parameter is not specified explicitly, we mean that the natural parameter is the parameter of the analysis.

Researchers in parameterized complexity have defined certain complexity classes of parameterized problems, containing problems that are not expected to have fixed-parameter tractable algorithms. One of the simplest of these is the class W[1]. Again, it has complete problems, and we can show that other parameterized problems are also W[1]-hard by translating them into these problems. If we can do this, it will provide strong evidence that the translated problem is not fixed-parameter tractable. Alternatively, we can sometimes show that a parameterized problem is hard by proving that it is NP-hard even for some fixed choice of the parameter's value. For instance, testing whether a graph can be colored with fewer than four colors is NP-hard, so parameterized graph coloring is unlikely to be fixed-parameter tractable.

If we are considering two parameters PARAMETER_1 and PARAMETER_2 such that $\text{PARAMETER}_1 \ll \text{PARAMETER}_2$, then it is better for an algorithm to be fixed-parameter tractable with respect to PARAMETER_1, for then it will automatically be fixed-parameter tractable with respect to PARAMETER_2. In this case we can think of PARAMETER_1 as "smaller" than PARAMETER_2, even though some of its actual values might be larger. But if $\text{PARAMETER}_2 \doteq \text{PARAMETER}_1$, then the two parameters are equivalent for fixed-parameter tractability. For, in this case, a problem is fixed-parameter tractable with respect to one of these parameters if and only if it is fixed-parameter tractable with respect to the other.

7.4 Complexity of Obstacle-Based Parameters

Let's apply these ideas to the complexity of monotone parameters of configurations. We saw in Observation 6.2 that one can test whether a configuration of size k is a subconfiguration of a larger configuration of size n, in time $O(n^k)$. Can we can solve this problem faster? As is often true in computational complexity, it is difficult to prove mathematically that the answer is no. However, the theorem below provides strong evidence that there may be nothing significantly better, such as an algorithm for which the exponent of n does not depend on k.

[7] Flum and Grohe (2006); Downey and Fellows (2013).

As with many of our theorems, we will defer its proof to a later chapter, in order to combine it with several related results using similar constructions.

Theorem 7.3

Testing whether configuration T is a subconfiguration of configuration S is NP-complete and W[1]-hard. If the respective sizes of T and S are k and n, then this problem has no algorithm with running time $n^{o(\sqrt{k})}$ unless the exponential time hypothesis is false.

Proof. We defer the proof to Section 15.3. □

The *exponential time hypothesis,* appearing in the statement of this theorem, is the assumption that no subexponential-time algorithm can find satisfying truth assignments to Boolean formulas.[8] Like the more famous question of whether all problems in NP can be solved in polynomial time, it remains unproven, but has been used as the basis of many hardness results.

In general, the parameters defined from obstacles can have quite high computational complexity.

Theorem 7.4

It is is Σ_2^{P}-complete to test whether AVOIDS$(C; S) \geq k$*, for an input consisting of configurations C and S and a number k. Testing whether* HITTING$(C; S) < k$ *(for an input of the same type, but with a different interpretation of k) is equivalent, and is therefore also Σ_2^{P}-complete.*

Proof. See Section 15.3. □

However, when we fix k to be a constant rather than treating it as an input variable, the two parameters AVOIDS$(C_1, C_2, \ldots)$ and HITTING$(C_1, C_2, \ldots)$ behave better.

Observation 7.5

For any set of obstacles $C_1, C_2, \ldots,$ AVOIDS$(C_1, C_2, \ldots)$ *has polynomial obstacle size.*

[8] Impagliazzo et al. (2001).

Proof. The obstacles to the property AVOIDS$(C_1, C_2, \ldots) < k$ are just the k-point configurations that have the property FORBIDDEN$(C_1, C_2, \ldots)$, so they all have size exactly k. □

For any fixed k, testing whether AVOIDS$(C_1, C_2, \ldots; S) < k$ may be done in polynomial time, by performing a brute force search among all size-k subconfigurations and then testing (using Observation 6.2) whether any of these subconfigurations avoids the given obstacles. Despite its high dependence on k, the time for this algorithm is polynomial in $|S|$. However, the degree of this polynomial (the exponent of $|S|$) depends on k, so the algorithm is not fixed-parameter tractable.

Open Problem 7.6

Is there a finite set of obstacles $C_1, C_2, \ldots$ for which AVOIDS$(C_1, C_2, \ldots)$ is not fixed-parameter tractable?

Theorem 7.7

There exists a parameter with polynomial obstacle size that is W[1]*-hard to compute.*

Proof. See Section 15.3. □

7.5 Parameterized Complexity of HITTING

We have stronger results on parameterized algorithms for HITTING$(C_1, C_2, \ldots)$, but proving them is trickier. It is convenient to begin with the following lemma.

Lemma 7.8 (The Erdős–Rado sunflower lemma (Erdős and Rado, 1960))

Let a and b be positive integers and let F be a family of $b!a^{b+1}$ finite sets, each of size at most b. Then F contains a subfamily of $a+1$ sets all of whose pairwise intersections are equal.

Figure 7.1 depicts a sunflower of seven sets ($a = 6$) with equal pairwise intersections. The proof of the following result follows Flum and Grohe (2006, section 9.1), which shows that the subfamily of the sunflower lemma can be

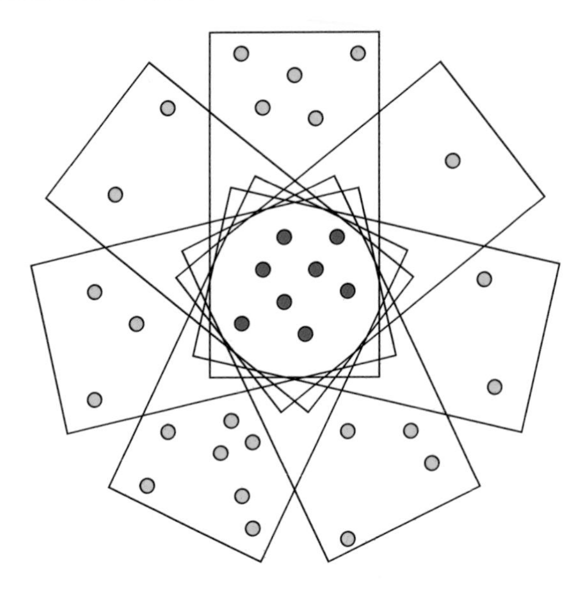

Figure 7.1. A sunflower of sets (shown as rectangles), all of whose pairwise intersections are equal.

found in polynomial time (when treating a as a variable and b as a constant). They use the sunflower lemma to solve the hitting set problem for families of sets of bounded size; this is the problem of choosing as few elements as possible so that each set in the given family includes at least one chosen element. In our applications of these results, the family of sets will be the instances of the obstacles used to define the HITTING parameter.

Theorem 7.9

The parameter HITTING$(C_1, C_2, \ldots)$ *has finite obstacle size if and only if* $C_1, C_2, \ldots$ *is a finite set. When this is the case, it has polynomial obstacle size, and computing* HITTING$(C_1, C_2, \ldots)$ *is fixed-parameter tractable.*

Proof. The obstacles $C_1, C_2, \ldots$ are also the obstacles for the property that HITTING$(C_1, C_2, \ldots) < 1$, so if there are infinitely many, then HITTING$(C_1, C_2, \ldots)$ does not have finite obstacles.

For the remainder of the proof, let $C_1, C_2, \ldots$ be finite, with each configuration in $C_1, C_2, \ldots$ having size at most b. We wish to bound the size of the obstacles for the property HITTING$(C_1, C_2, \ldots) < k$. For any given configuration S that does not have this property, let F be the family of instances of the subconfigurations $C_1, C_2, \ldots$ in S. We will remove instances one at a time from F, preserving the property that the remaining family of instances cannot be hit by fewer than

k points of S, until no more removals are possible. Once this happens, S will necessarily contain an obstacle that is a subconfiguration of the union of the remaining instances in F.

To choose instances to remove from F, while $|F| \geq b!k^{b+1}$ we use the sunflower lemma to find a collection of $k+1$ instances in F, all pairs of which have the same intersection I. We then remove from F any one of the instances in the collection. Any hitting set for the remaining instances in F with fewer than k points must include one of the points in I, for otherwise it could not hit all the remaining instances in the collection. This point of I would also hit the removed instance. Therefore, removing an instance from F in this way cannot change the existence or nonexistence of a hitting set of size less than k.

Once no more sunflowers can be found, the remaining family F must include fewer than $b!k^{b+1} = O(k^{b+1})$ instances, and its union (within which an obstacle can be found) also has size $O(k^{b+1})$, polynomial in k as b is a fixed constant. Therefore, HITTING$(C_1, C_2, \ldots)$ has polynomial obstacle size.

It remains to prove that computing HITTING$(C_1, C_2, \ldots)$ is fixed-parameter tractable. A kernelization algorithm can follow the same general outline given above, performing the following steps.

Algorithm 7.10 (Kernelization for HITTING)

---

1. Set $b = \max_i |C_i|$, the largest size of any of the obstacles.
2. Loop through the numbers $k = 1, 2, 3 \ldots$. For each choice of k, perform the following steps.
   a. Find all instances of $C_1, C_2, \ldots$ in the input configuration. Let F be the collection of instances found in this way.
   b. While F includes at least $b!k^{b+1}$ instances, use the sunflower lemma to find $k+1$ instances whose pairwise intersections are all equal. Choose an arbitrary instance from this sunflower and remove it from F. This removal is safe, because any hitting set of size less than k for the remaining instances in the sunflower would also hit the removed instance.
   c. After F has been reduced to fewer than $b!k^{b+1}$ instances, let U be the subconfiguration consisting of the points in the remaining instances of F.
   d. Use brute force search to test whether U has a hitting set of size less than k. If this search succeeds, stop the algorithm and return this hitting set as the result. □

An alternative fixed-parameter-tractable kernelization algorithm for the hitting set problem, which yields somewhat smaller kernels and could also be applied to the problem here, is described by Abu-Khzam (2010).

We have not stated results here about the complexity of computing the parameter PARTITION$(C_1, C_2, \ldots)$. But we will see in Theorem 10.16 that, even for a finite set of obstacles and a fixed parameter value, PARTITION$(C_1, C_2, \ldots)$

can be NP-hard and have infinitely many obstacles. Although we saw that configuration size can be defined using any of AVOIDS, HITTING, or PARTITION (Example 5.13), this kind of equivalence is not true more generally. Because PARTITION can be so much harder than AVOIDS and HITTING, some parameters that can be defined using PARTITION cannot be redefined to use AVOIDS or HITTING.

8 Collinearity

8.1 Planting Orchards

According to the *Sylvester–Gallai theorem,* every configuration of more than one point that is not of the form LINE(n) contains an *ordinary line,* a line through exactly two of the points.[1] We call the other lines, through three or more points, *heavy lines.* The number of heavy lines is a monotone parameter that we denote HEAVY-LINES(S). It has been investigated since at least the early nineteenth century in the *orchard-planting problem.*

Open Problem 8.1 (the orchard-planting problem)

For each number n, what is the maximum possible value of HEAVY-LINES(S) among all configurations S of size n?

The name comes from the fact that the trees in an orchard are typically planted in rows. One can interpret the points of a configuration as positions to plant trees within an orchard, and the heavy lines of the configuration as the rows of trees that would result from planting trees in these positions. Figure 8.1 shows the optimal solution for $n = 11$, with 16 lines.

Jackson (1821) posed several variations of the orchard-planting problem, in a sequence of verses. His first verse asks for a nine-point configuration with ten heavy lines, the largest number possible:

[1] For more on this problem, see Borwein and Moser (1990) and Pach and Sharir (2009).

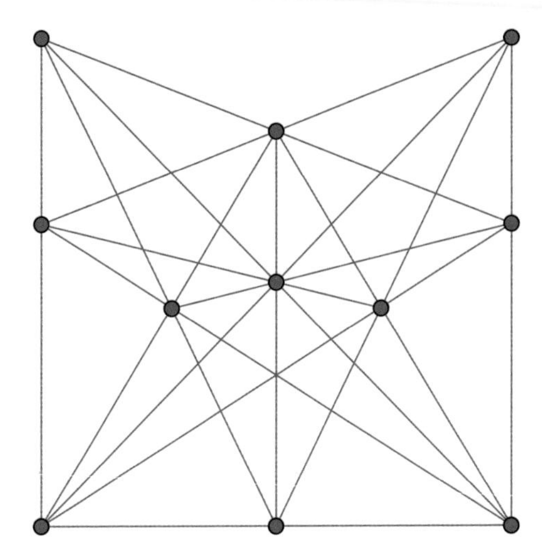

Figure 8.1. An orchard of 11 trees (green), watered in sets of three by 16 linear irrigation channels (blue). This configuration gives the maximum number of heavy lines for any configuration of 11 points. The two trees on the left and right sides of the orchard divide their sides in the golden ratio, determining the positions of the remaining trees. Redrawn from figure 1 of Burr et al. (1974).

Your aid I want, nine trees to plant
in rows just half a score;
And let there be in each row three.
Solve this: I ask no more.

We have already seen the answer to this riddle, in Figure 6.1. But for 15 points Jackson asks only for 23 heavy lines, when at least 31 are possible. And for 17 points he asks only for 24 heavy lines, when at least 40 are possible.[2] Loyd (1914) repeated the nine-point version of the question in a different format, with eggs in place of trees. Figure 8.1 shows the optimal solution for this problem when there are eleven trees or eggs.

Each heavy line, in any configuration, is determined by at least three pairs of points. Those same pairs cannot determine any other line, and each n-point configuration has only $\binom{n}{2}$ pairs of points to go around. Therefore, the number of heavy lines in an n-point configuration can be at most

$$\left\lfloor \binom{n}{2} \Big/ 3 \right\rfloor = \left\lfloor \frac{n(n-1)}{6} \right\rfloor.$$

However, this bound is not tight. Not all of the pairs of points can contribute to this count of heavy lines, because some pairs form ordinary lines instead (the Sylvester–Gallai theorem again). Each ordinary line subtracts one from the number of pairs that form heavy lines, so when more ordinary lines are

[2] For the known values of the maximum number of heavy lines, see sequence A003035 in the Online Encyclopedia of Integer Sequences (OEIS Foundation, 2017).

included in a configuration, it can have fewer heavy lines. In this way, the orchard-planting problem is closely related to the question of how many ordinary lines a set of noncollinear points must have.[3] Kelly and Moser (1958) proved that n noncollinear points have at least $3n/7$ ordinary lines, and Csima and Sawyer (1993) improved this to $6n/13$ (except when $n = 7$). These bounds in turn lead to tighter bounds on the number of heavy lines.

Burr et al. (1974) found a construction based on elliptic curves for configurations with many heavy lines, related to an earlier construction of J. J. Sylvester. Their construction produces sets of points S with $|S| = n$ and with

$$\text{HEAVY-LINES}(S) = \left\lfloor \frac{n(n-3)}{6} \right\rfloor + 1.$$

This expression matches the upper bounds above in its leading term ($n^2/6$), but differs from them by a lower-order term proportional to n.

Green and Tao (2013) gave a tight $n/2$ bound on how few ordinary lines there can be, for all sufficiently large values of n. Using this to bound the number of heavy lines is, again, not tight. However, in the same paper, Greene and Tao also showed that for all sufficiently large n, the construction of Burr et al. gives the exact maximum for the number of heavy lines. Thus, the orchard-planting problem remains open only for small values of n. The smallest value of n for which an exact solution remains unknown is $n = 15$.[4] Although the result of Green and Tao provides an exact solution for large n, the point at which n becomes large enough is also unknown.

8.2 Definitions

As well as HEAVY-LINES, many other monotone parameters involve lines of points. A configuration lies on a single line if and only if it does not contain a triangle, the configuration POLYGON(3). So, when we use the triangle as an obstacle, the properties and parameters that we obtain from it via Definition 5.8 and Definition 5.12 will be related to the property of being on a line.

Definition 8.2

We define:

$$\text{COLLINEAR} = \text{FORBIDDEN}\big(\text{POLYGON}(3)\big)$$
$$\text{ONLINE} = \text{AVOIDS}\big(\text{POLYGON}(3)\big)$$
$$\text{OFFLINE} = \text{HITTING}\big(\text{POLYGON}(3)\big)$$
$$\text{LINE-COVER} = \text{PARTITION}\big(\text{POLYGON}(3)\big).$$

[3] Burr et al. (1974). [4] OEIS Foundation (2017).

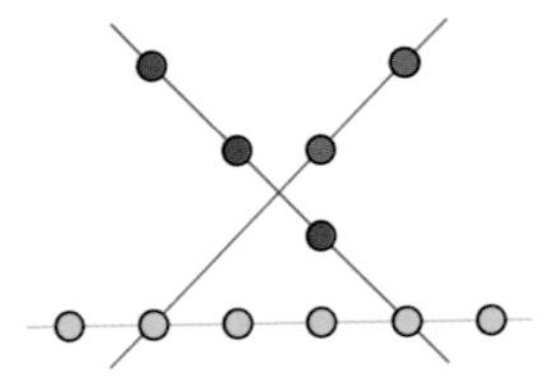

Figure 8.2. A configuration S for which COLLINEAR(S) is false (the points are not all on one line), ONLINE(S) $= 6$ (six points lie on the yellow line), OFFLINE(S) $= 5$ (removing the five red and blue points leaves a collinear subconfiguration), and LINE-COVER $= 3$ (the yellow, blue, and red lines cover all points of S, and they cannot be covered by only two lines).

Here, the property COLLINEAR is true when all points are on one line. The parameter ONLINE gives the maximum number of points on a single line of the given configuration. The parameter OFFLINE gives the minimum number of points to remove from the configuration to leave the remaining points collinear. And the parameter LINE-COVER gives the smallest number of collinear subconfigurations into which the configuration may be partitioned. These properties and parameters are illustrated in Figure 8.2.

8.3 Inequalities

Observation 8.3

LINE-COVER $\ll$ OFFLINE.

Proof. This follows from the inequality, valid for all configurations S, that

$$\text{LINE-COVER}(S) \leq 1 + \left\lceil \frac{\text{OFFLINE}(S)}{2} \right\rceil.$$

If S can be reduced to points on a line by the removal of OFFLINE(S) points, it can be covered by that one line together with $\lceil \text{OFFLINE}(S)/2 \rceil$ additional lines that cover the removed points in pairs (Figure 8.3, left). This inequality is a special case of Observation 5.15, the inequality between PARTITION and HITTING.

To complete the proof we need configurations with bounded LINE-COVER and unbounded OFFLINE. GRID$(2, n)$ works: it has LINE-COVER$\big($GRID$(2, n)\big) = 2$ but OFFLINE$\big($GRID$(2, n)\big) = n$ when $n > 1$. Figure 8.3 (right) shows an example. □

Observation 8.4

HEAVY-LINES $\ll$ OFFLINE.

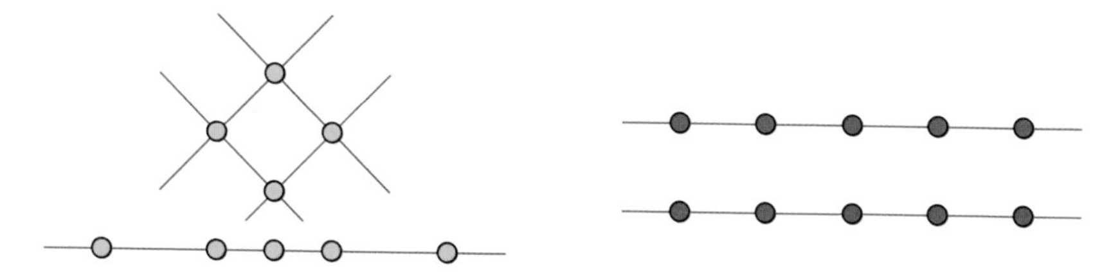

Figure 8.3. Illustration for Observation 8.3. Left: any configuration S can be covered by $1 + \lceil \text{OFFLINE}(S)/2 \rceil$ lines, by using one line through as many points as possible and then covering the remaining $\text{OFFLINE}(S)$ points in pairs. Right: when $S = \text{GRID}(2, n)$, $\text{LINE-COVER}(S) = 2$ but $\text{OFFLINE}(S) = n$, showing that these two parameters can be far apart.

Proof. If one line L covers all but k points, then every other heavy line uses at least two of those k points, so the number of heavy lines is at most $\binom{k}{2} + 1$. For configurations with bounded values of HEAVY-LINES and unbounded values of OFFLINE, consider either $\text{GRID}(2, n)$ again or $\text{POLYGON}(n)$. □

In Observation 17.23 and Observation 17.24, we will refine Observation 8.3 and Observation 8.4 by sandwiching another parameter between LINE-COVER, HEAVY-LINES, and OFFLINE.

8.4 Finding All Lines

Collinearity may easily be tested in linear time. To do so, choose an arbitrary pair of points and check that the remaining points are all collinear with them.

One way to compute $\text{ONLINE}(S)$ and $\text{OFFLINE}(S)$ would be to list the subsets of points on each line of a given configuration, and then from that list determine which line has the most points. We may use sorting to do this in time $O(n^2 \log n)$ and space $O(n)$, according to the following steps.

Algorithm 8.5 (Sorting-based search for lines with many points)

1. Loop over the points of the given configuration S, and for each point p perform the following steps.
   a. Sort all of the points of S other than p, using the slope of line pq as the sort key for each point q. Once the points are sorted, collect them into subsets of points that all have the same slope.
   b. Each subset of points collected in this way defines a line, but possibly one that might have been already output. Check whether p is the first point of this line (that is, the earliest one to be considered in the outer loop of the algorithm) and, if so, output this subset as another line.

2. Among all of the sets of points generated in the previous loop, let L be the one with the most points. Compute ONLINE$(S) = |L|$ and OFFLINE$(S) = |S| - |L|$.

If the points of S are already represented by integer Cartesian coordinates, the slope of any line determined by two points (x_1, y_1) and (x_2, y_2) may be computed directly as

$$\frac{y_1 - y_2}{x_1 - x_2},$$

or as a special nonnumerical value ∞ when $x_1 = x_2$ and the two points define a vertical line. Alternatively, for points with integer coordinates, grouping points into subsets with the same slope may be performed even more quickly by using a hash table with the slope as the hash key.

On the other hand, if the configuration can only be accessed through queries to orientations of triples of points, the algorithm outlined above can nevertheless be performed efficiently. In this case, we have no concept of vertical, but for every point p we can choose one other point q (arbitrarily) as the one that defines a vertical line with p. Then, for this choice of vertical, we can compare the slopes of any two lines pr and ps by testing the orientations of the triples (p, q, r) and (p, q, s) to determine whether r and s are left or right of the vertical line through p. If r and s are on the same side, the ordering between the slopes of lines pr and ps is given by the orientation of triple (p, r, s), while if r and s are on opposite sides, we should negate this orientation to give the ordering. This method for comparison allows us to use any comparison-based sorting subroutine in Algorithm 8.5.

Because sorting can be done in time $O(n \log n)$, and this algorithm performs n sorts, its time is $O(n^2 \log n)$. When all lines are to be output, this time bound is nearly optimal, because it would not be possible to run faster than the output size, which can be $\Theta(n^2)$. The logarithmic factor between the upper and lower bounds can be eliminated, even in a comparison-based model of computation (where using hash tables instead of sorting is not allowed) at the expense of using a more complicated algorithm.

Theorem 8.6 (Edelsbrunner and Guibas, 1989)

The set of lines generated by a given configuration of points, and the subsets of points on each line, can be found in time $O(n^2)$ and space $O(n)$. Therefore, both ONLINE(S) *and* OFFLINE(S) *can also be computed in time $O(n^2)$ and space $O(n)$.*

This result is a corollary of a more general technique of Edelsbrunner and Guibas for processing arrangements of lines by a technique that they call *topological sweeping*, in the same time and space bounds. Edelsbrunner and Guibas use the theory of projective duality to transform a given set of points into a dual set of lines in the projective plane. This transformation has the property that any subset of points through a common line in the original input corresponds to a subset of lines through a common point in the dual problem, and vice versa. Then, by using topological sweeping to find all crossing points of the dual line arrangement, they can also find all collinear sets of points in the original problem.

This line-finding algorithm can be used to provide a faster algorithmic solution to the puzzle of Figure 6.1.

Corollary 8.7

We can count the triangles in a given configuration in time $O(n^2)$.

Proof. Let the size of the configuration be n, and let ℓ_i denote the number of i-point lines. Then the number of triangles can be calculated as

$$\binom{n}{3} - \sum_{i=3}^{n} \ell_i \binom{i}{3}.$$ □

The $\Omega(n^2)$ output-size lower bound does not apply to the problem of computing ONLINE, because that problem's output is too small (just a single number). However, some research suggests that quadratic time is the best possible, even for testing whether ONLINE$(S) = 2$ (that is, whether the input is in general position, or whether it has $\binom{n}{3}$ triangles). This problem has $\Omega(n^2)$ lower bounds in some models of computation that can access the input only through a limited set of predicates.[5] Its complexity continues to be studied using a more abstract version of the problem, defined below.[6]

Definition 8.8

In the *3SUM* problem, the input is three lists A, B, and C of n numbers each, and the goal is to select one number from each list so the three selected numbers add to zero. A problem is 3SUM-hard if the 3SUM problem can be translated to an instance of the given problem of size $O(n)$, so that an algorithm for the given problem would also lead to an algorithm with the same asymptotic runtime for 3SUM.

[5] Erickson and Seidel (1995); Ailon and Chazelle (2005).
[6] Gajentaan and Overmars (1995).

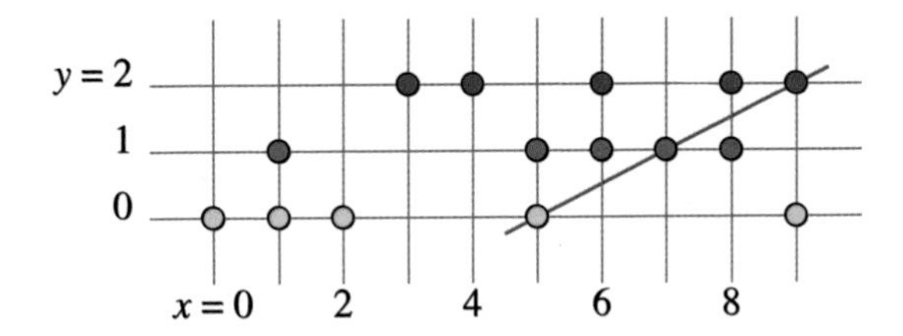

Figure 8.4. Translation of 3SUM into collinearity. These points come from the 3SUM instance $A = \{0, 1, 2, 5, 9\}$ (the bottom line of yellow points), $B = \{-2, -10, -12, -14, -16\}$ (blue points), and $C = \{3, 4, 6, 8, 9\}$ (red points). The red non-horizontal line through three of these points corresponds to the 3SUM solution $5 + (-14) + 9 = 0$. Other solutions are not shown.

For instance, it is 3SUM-hard, for configurations S with LINE-COVER(S) $= 3$, to determine whether there is a fourth line (different from the three covering lines) through at least three points. To see this, use the following translation:[7]

- For each $a_i \in A$ create a point with Cartesian coordinates $(a_i, 0)$.
- For each $b_j \in B$ create a point with Cartesian coordinates $(-b_j/2, 1)$.
- For each $c_k \in C$ create a point with Cartesian coordinates $(c_k, 2)$.

Then $a_i + b_j + c_k = 0$ if and only if the corresponding three points are collinear. Figure 8.4 depicts an example. An alternative transformation avoids the three covering lines. Choose a large number N, and map

$$
\begin{aligned}
a_i &\mapsto (a_i + N, (a_i + N)^3) \\
b_j &\mapsto (b_j + 2N, (b_j + 2N)^3) \\
c_k &\mapsto (c_k - 3N, (c_k - 3N)^3)
\end{aligned}
$$

for each $a_i \in A$, $b_j \in B$, and $c_k \in C$. N should be large enough to ensure that the only triples of x-coordinates that sum to zero come from triples of points that include one point from each of A, B, and C. Although this transformation is more difficult to draw, it has the same property of mapping 3SUM solutions one-to-one to collinear triples of points. Therefore, it is 3SUM-hard to test whether a configuration is in general position.[8]

Slightly subquadratic algorithms are now known for 3SUM.[9] Their time bounds also apply to line detection for points on three parallel lines. However, the topological sweeping algorithm of Theorem 8.6 remains the fastest one known for detecting lines of three or more points in the general case.

Computing ONLINE and OFFLINE is hardest when ONLINE(S) is small. If, instead, there exists a line through many points, the problem can be solved more quickly: faster by nearly a factor of the number of points on the line. For this

[7] Gajentaan and Overmars (1995). [8] Gajentaan and Overmars (1995).
[9] Baran et al. (2008); Grønlund and Pettie (2014); Chan and Lewenstein (2015).

case, Guibas et al. (1996) show how to compute $\textsc{online}(S)$ in time

$$O\left(\frac{|S|^2}{\textsc{online}(S)}\log\frac{|S|}{\textsc{online}(S)}\right).$$

8.5 Complexity of Line Covers

Computing or approximating $\textsc{line-cover}(S)$ has been studied for its own sake, but it also forms a limiting case of *projective clustering*, the problem of covering a configuration S with a small number of thin strips rather than lines.[10] It has higher complexity than the other collinearity-based parameters.

Theorem 8.9 (Megiddo and Tamir, 1982)

It is NP*-complete to determine whether* $\textsc{line-cover}(S) < k$*, when* S *and* k *are both inputs to the problem.*

Despite its NP-completeness, LINE-COVER can still be computed efficiently when the number of lines is small.

Theorem 8.10 (Langerman and Morin, 2005)

Computing LINE-COVER *is fixed-parameter tractable.*

Proof. We sketch here a simple kernelization that proves Theorem 8.10. An algorithm for testing whether $\textsc{line-cover}(S) < k$ may be obtained by following the steps below.

Algorithm 8.11 (Kernelization for LINE-COVER)

1. Initialize a set K of points and another set F of lines, both empty. K will eventually be the kernel that we use, and F will be a set of lines that are forced to be part of any cover by fewer than k lines.
2. For each point p of S, in an arbitrary order, perform the following steps.
   a. Check whether p belongs to any lines of F, and if so discard p and continue to the next point of S, skipping the following steps.
   b. Add p to K.

[10] Agarwal and Procopiuc (2003).

c. While K contains a line L that passes through $k-|F|$ points, add this line to F and remove from K the points that it contains. L must be part of any cover, because covering all of its points with separate lines would be too expensive.
d. If, after the previous loop, $|K| \geq (k-|F|)^2$, then there can be no cover by fewer than k lines, because K can only be covered by lines through fewer than $k-|F|$ points and there would be too many such lines in any cover. In this case, halt the algorithm and return that LINE-COVER$(S) \geq k$.

3. If the algorithm processes all points without halting, apply a brute force search to test whether K can be covered with fewer than $k-|F|$ lines.

Each step except the last is polynomial, and the last step applies a brute force search to a kernel of size less than k^2. Therefore, this algorithm is fixed-parameter tractable. □

The dependence of the kernel size on k is necessary: under standard assumptions of complexity theory, no polynomial-time kernelization has a kernel of size $O(k^c)$ for c less than two.[11] Faster fixed-parameter algorithms than the one above are known; the fastest known takes time $O\big(n \log k + (k/1.35)^k\big)$, for configurations of size n.[12]

8.6 Property Testing

A property testing algorithm that distinguishes configurations with small line covers from configurations that are ϵ-far from having a small line cover can use significantly smaller samples than the ones of Theorem 6.8. Indeed, for this problem, the sample size can be independent of the input size.

Theorem 8.12

For any constants k, ϵ, and p with $0 < \epsilon < 1$ and $0 < p < 1$, sample-based property testing can distinguish inputs with LINE-COVER $< k$ *from inputs ϵ-far from this property, with sample size $O(1)$ and false positive rate p.*

Proof. What we need to prove is that, if S is ϵ-far from having LINE-COVER $< k$, and we choose a random sample of a large enough but constant size from S, then the resulting subconfiguration probably has LINE-COVER $\geq k$. What we will actually prove is that if we keep choosing points randomly one-by-one from

[11] Kratsch et al. (2016). For additional research on fixed-parameter tractable algorithms for this problem, see Grantson and Levcopoulos (2006), Estivill-Castro et al. (2009), and Wang et al. (2010).

[12] Wang et al. (2010).

S, until reaching a subconfiguration that has LINE-COVER $\geq k$, the size of the subconfiguration will on the average be constant. These two claims are related by *Markov's inequality*, which states that any nonnegative random variable is unlikely to be much bigger than its expected value. We apply this inequality to a random variable that counts the number of points at which the one-by-one sample first has LINE-COVER $\geq k$. By Markov's inequality, a sample that is bigger than the expected value of this random variable by a factor of $1/p$ will also have LINE-COVER $\geq k$, with probability at least $1 - p$, as needed by the property testing algorithm. This use of Markov's inequality allows us to ignore p for the rest of the proof, concentrating only on ϵ and k.

We prove that the expected size of the one-by-one sample is $O(1)$ by induction on k. As a base case, when $k = 1$, the expected sample size needed until we reach LINE-COVER $\geq k$ is 1. For, it takes one line to cover one point.

Now, suppose that $k > 1$, and consider sampling points from S one-by-one as described above. This sampling process generates a sequence of subconfigurations S_i with $|S_i| = i$, generated by adding a single point to S_{i-1}. At some step i, we will reach a subconfiguration S_i with LINE-COVER $\geq k - 1$. By the induction hypothesis, the expected size i of S_i is $O(1)$.

If $|S_i| \geq \epsilon|S|/2$, we may replace S_i by S, producing a configuration with LINE-COVER $\geq k$ because S is assumed to be ϵ-far from having this property. This replacement only increases the size of the sample by a constant factor, $2/\epsilon$. But if not, then S_i is small enough that $S \setminus S_i$ is still $\epsilon/2$-far from having LINE-COVER $\geq k$. In this case, continue the one-by-one sampling process until we reach a subconfiguration S_j such that $S_j \setminus S_i$ has LINE-COVER $\geq k - 1$. By the induction hypothesis, the expected size j of S_j is still $O(1)$.

If S_j already has LINE-COVER $\geq k$, we are done. Otherwise, we have disjoint subconfigurations S_i and $S_j \setminus S_i$ with LINE-COVER $\geq k - 1$, which we can reduce to obstacles X_i and X_j of guaranteed size $O(1)$ (Theorem 8.14, below). Let $C_1, C_2, \ldots$ be the family of $(k-1)$-line covers of S_j. Each cover C_h consists of $k - 1$ lines, each having a point in X_i and a point in X_j. There are $O(1)$ lines to choose from, so there are $O(1)$ covers C_h. Each C_h fails to cover at least $\epsilon|S|$ points in S, none of which has been sampled already. Therefore, each point that we sample after S_j has a constant probability of being uncovered by C_h, preventing C_h from covering the whole sample set. After an additional $O(1)$ expected sample points, all of the covers C_h will be eliminated in this way, and the sample will have LINE-COVER $\geq k$. □

8.7 Obstacle Size and Well-Quasi-Ordering

Both ONLINE(S) and OFFLINE(S) have polynomial obstacle size, by Observation 7.5 and Theorem 7.9. We can show this more directly, improving the bound on obstacle size for OFFLINE(S), as follows. Any configuration S with OFFLINE $\geq k$ must contain a small forbidden subconfiguration T of size at most $2k + 1$, found according to the following case analysis.

- If $\textsc{online}(S) > k$, let T consist of $k+1$ points on some line L and any k points not on L. Then removing any $k-1$ points from T leaves at least two points on L and one point off it, so $\textsc{offline}(T) \geq k$.
- Otherwise, $\textsc{online}(S) = k$. Let T be any $2k$ points of S. Removing $k-1$ of them will leave $k+1$, too many to be collinear. So, again $\textsc{offline}(T) \geq k$.

T might not be an obstacle, but it contains an obstacle of size at most $2k+1$.

Theorem 8.13

For any fixed k, the configurations S with $\textsc{offline}(S) < k$ *are well-quasi-ordered. The number of such configurations with n points is bounded by a polynomial in n. Any monotone parameter is nonuniform fixed-parameter tractable when restricted to these configurations.*

Proof. We defer the proof to Section 17.6. □

Theorem 8.14

The LINE-COVER *parameter has polynomial obstacle size.*

Proof. This follows from the kernelization for LINE-COVER in Algorithm 8.11. To find an obstacle for $\textsc{line-cover}(S) < k$, apply the following case analysis.

- If S has a line L with k or more points on it, then L must be part of any cover with fewer than k lines, for if it were not included, then each of these points would need to be covered by a separate line, causing the cover to have too many lines. Therefore, in this case, one can eliminate L and all points on L, and reduce the parameter by one.
- Otherwise, if all lines have fewer than $k-1$ points, then at most $(k-1)^2$ points can remain if all are to be covered by only $k-1$ lines. So, any subconfiguration of $(k-1)^2+1$ points has $\textsc{line-cover} \geq k$.

Therefore, a small forbidden subconfiguration can be found within S either as k points on some line L plus a recursively constructed obstacle for one fewer line in the subconfiguration of points not on L or as any subconfiguration of $(k-1)^2+1$ points with no k-point line. Its maximum size is the greater of two quantities: k plus the obstacle size for $k-1$ in the first case or $(k-1)^2+1$ in the second case. For $k=2$ or $k=3$ we get three or six points, respectively, and for all larger k we get the $(k-1)^2+1$ size from the second case. □

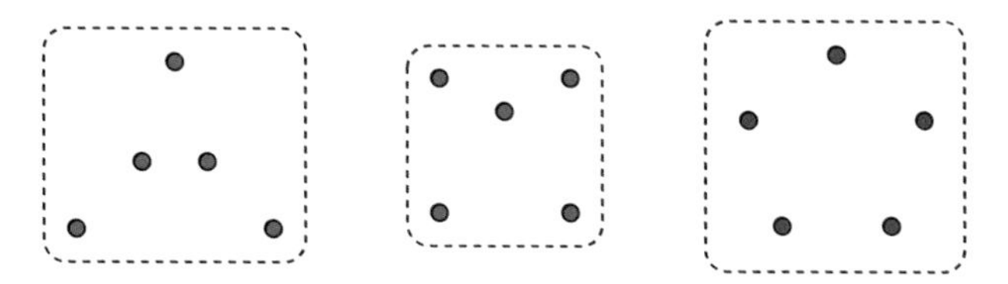

Figure 8.5. The three configurations with five points in general position, obstacles for LINE-COVER(S) < 3 and OFFLINE(S) < 3 (Example 8.15).

Example 8.15

We can list the obstacles for LINE-COVER(S) < 3 and for OFFLINE(S) < 3 as follows. We begin with LINE-COVER(S) < 3, as it has fewer obstacles. By the proof outline above, each obstacle either has three points in a line and three points in general position off the line (for a total of six points) or five points in general position.

To be minimal, the six-point sets of this form must not contain five points in general position. They can have no four-point lines, for if they did the other two points could be covered by another line, giving a two-line cover. Each two three-point lines share a point, for otherwise they would form a two-line cover. There must be three distinct three-point lines, not all through the same point, for otherwise all three-point lines could be eliminated by removing one point, leaving five points in general position. The only remaining possibility is a triangle of three three-point lines, with a point at each vertex of the triangle where two lines cross, and with another point somewhere else on each line. The third point on each line can be placed in three different ways:

1. outside the triangle, clockwise of it along the line through one of its sides
2. outside the triangle, and counterclockwise of it, or
3. between the two triangle vertices.

Each obstacle of this type can be described by making one of these three choices for each of the three triangle sides, but cyclic permutations of these choices produce the same configuration. Therefore, there are eleven different six-point obstacles, described by the eleven choice sequences 111, 112, 113, 122, 123, 132, 133, 222, 223, 233, and 333. (Some of these are mirror images of each other but they are nevertheless distinct as configurations.)

Four of these sequences (112, 122, 133, and 233) have a triple of points whose orientation is not determined by the sequence. If this triple is collinear, the points form a complete quadrilateral, while otherwise each of these four sequences gives rise to two different configurations. Thus, there are 16 six-point obstacles, not just 11, as shown in Figure 8.6. Together with the three different configurations of five points in general position (shown in Figure 8.5), this gives 19 obstacles in total for LINE-COVER(S) < 3.

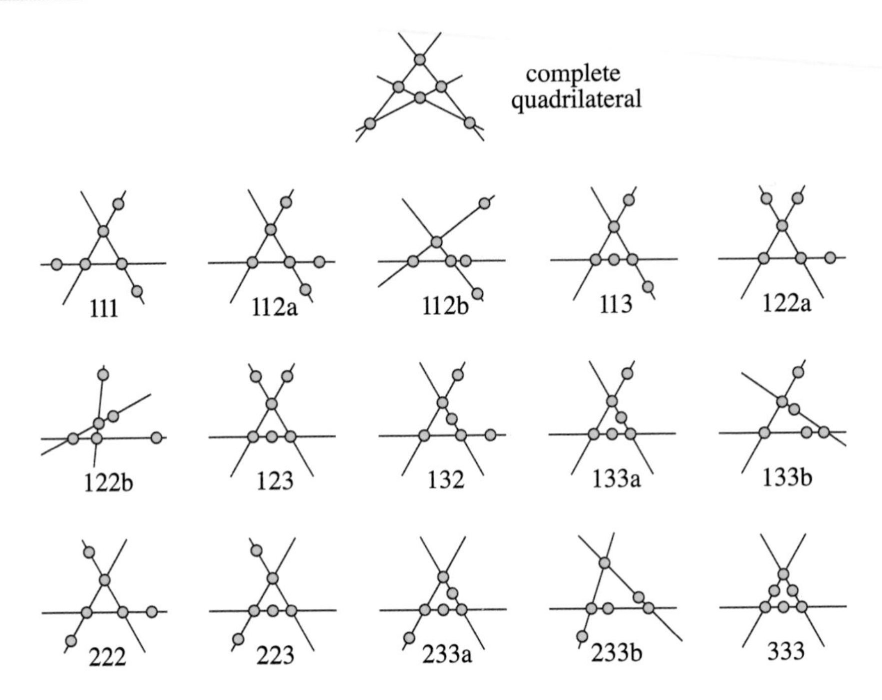

Figure 8.6. The 16 configurations that have six points in three lines of three points. They are obstacles for LINE-COVER$(S) < 3$, OFFLINE$(S) < 3$ (Example 8.15), and DELETE-TO-GENERAL$(S) < 2$ (Example 9.4).

By almost the same reasoning, these 19 configurations are also all obstacles for OFFLINE$(S) < 3$. Each requires the removal of three points to make the remaining points collinear, and is minimal with this property. However, there are also four more obstacles for OFFLINE$(S) < 3$, the four configurations that have six points in two disjoint lines of three (shown in Figure 8.7).

Theorem 8.16

The configurations S with LINE-COVER$(S) \leq 2$ *are well-quasi-ordered. Every monotone parameter is nonuniform fixed-parameter tractable when restricted to these configurations. However, the configurations S with* LINE-COVER$(S) \leq 3$ *are not well-quasi-ordered. There are $2^{\Theta(n \log n)}$ of these configurations with n points.*

Proof. For the results for LINE-COVER ≤ 2, see Section 17.6. For the results for LINE-COVER ≤ 3, see Section 15.4. □

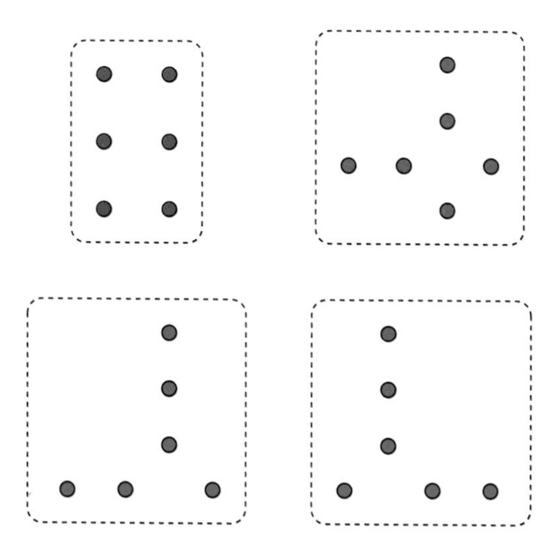

Figure 8.7. The four configurations in which six points form two disjoint lines of three points. These are obstacles for OFFLINE(S) < 3 (Example 8.15) and for DELETE-TO-GENERAL(S) < 2 (Example 9.4).

8.8 Approximation

Theorem 8.17 (Dumitrescu and Jiang, 2015)

Finding an approximation to LINE-COVER(S) *is* APX*-hard, but it may be approximated to within a logarithmic approximation ratio.*

This $O(\log n)$ approximation ratio applies more generally to any partition problem where we can find the largest obstacle-avoiding subconfiguration efficiently. To approximate the minimum line cover with this ratio, repeatedly find a line through as many not-yet-covered points as possible using Algorithm 8.18 (see Figure 8.8). An algorithm like this one that repeatedly makes the largest choice that it can is called a *greedy algorithm.*

Algorithm 8.18 (Greedy line cover)

1. Initialize L to an empty set of lines.
2. While the set U of input points not covered by L is nonempty, do the following steps.
   a. Find a line ℓ through as many points of U as possible.
   b. Add ℓ to U.
3. Return L as the approximate line cover.

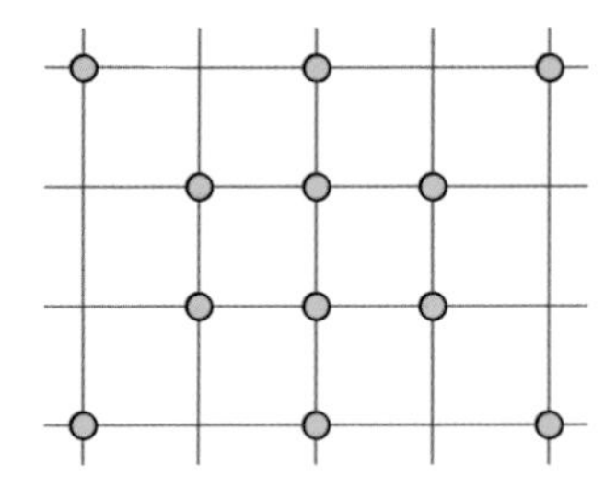

Figure 8.8. A greedy approximation to LINE-COVER for this configuration chooses the central four-point line first, and then can only cover the remaining points in pairs, producing the five vertical red lines. However, there is a better solution using only the four horizontal blue lines.

If the optimal cover uses k lines, then each greedily chosen line will cover at least a $1/k$ fraction of the remaining uncovered points. Repeatedly reducing the number of uncovered points by a $1/k$ fraction must reduce this number from n to less than one in at most $k \ln n$ steps, giving an approximation ratio of $\ln n$.

Example 8.19

We have already seen in Figure 8.8 that the greedy line-covering algorithm can be tricked into finding a nonoptimal line cover. How bad can it get?

The configuration illustrated in Figure 8.9 shows that in fact the greedy line-covering algorithm can do no better than the logarithmic approximation

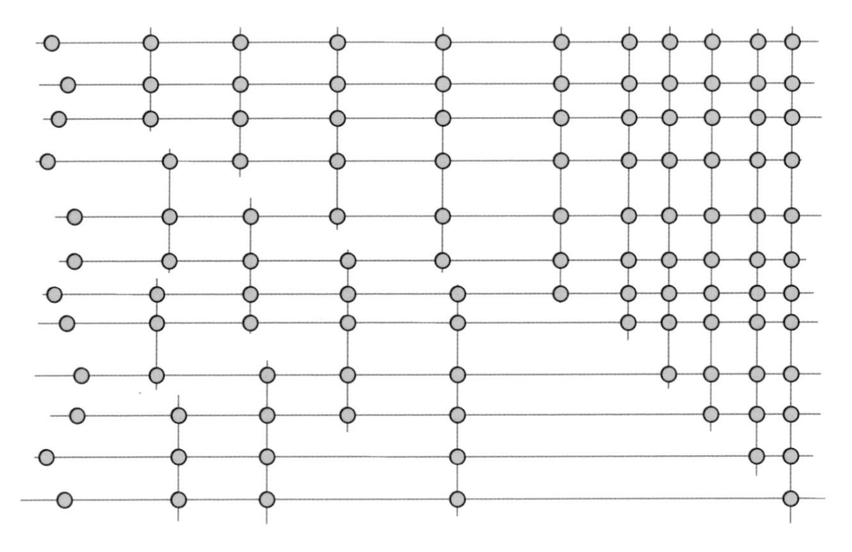

Figure 8.9. A configuration that causes the greedy line-covering approximation to perform badly (Example 8.19).

ratio proven above. This configuration is constructed from some number k of horizontal lines (in the figure, $k = 12$), and additional vertical line segments through these lines. For each integer i in the range from 3 to k, we include $\lfloor k/i \rfloor$ vertical line segments that pass through disjoint groups of i consecutive horizontal lines. For instance, for $k = 12$ and $i = 3$, there are four vertical line segments (the leftmost four in the figure) that pass through the horizontal lines numbered (from top to bottom) 1–3, 4–6, 7–9, and 10–12. We place a point on each intersection of a horizontal line and a vertical segment, and one more point on each horizontal line. These lines, line segments, and points are spaced irregularly so that the only lines that contain three or more points are the horizontal and vertical ones of the construction.

On this configuration, the greedy algorithm covers the points using the vertical line segments. It chooses first the vertical segment through k points (rightmost in the figure), because each horizontal line or other vertical segment has at most $k-1$ points. Next, it chooses the vertical segment that passes through $k-1$ points, because the horizontal lines (and all other vertical segments) have at most $k-2$ uncovered points on them. It continues in this way (from right to left in the figure), at each step choosing the vertical segment through the most remaining points. When it has chosen all such segments, there will remain k points to cover, the leftmost ones in the figure, which it does using $\lceil k/2 \rceil$ lines (chosen arbitrarily) to cover two points per line. Thus, the greedy algorithm ends up choosing

$$\left\lceil \frac{k}{2} \right\rceil + \left\lfloor \frac{k}{3} \right\rfloor + \left\lfloor \frac{k}{4} \right\rfloor + \cdots = k \ln k - O(k)$$

lines in its cover. On the other hand, the optimal cover uses only the k horizontal lines. Therefore, the approximation ratio of the greedy algorithm on this example is $\ln k - O(1)$. The number n of points in the configuration is $O(k^2)$, so we can rewrite the approximation ratio in terms of n as $\left(\frac{1}{2} - o(1)\right) \ln n$. This lower bound on the approximation ratio matches to within a constant factor the $\ln n$ upper bound for the algorithm.

An alternative algorithm for a wide class of geometric set cover problems also provides a logarithmic approximation to LINE-COVER. The constant factor in its approximation ratio is larger. But in exchange the algorithm is more flexible in ways that (if an open problem could be solved) might cause it to have a better-than-logarithmic approximation ratio.

This algorithm is based on a geometric structure known as an *ϵ-net*. To define the type of ϵ-net that we need in the approximation for LINE-COVER, consider the family of all the lines through pairs of points of a given configuration. We will assign weights to these lines, arbitrary positive real numbers. A point of the configuration is defined to be ϵ-heavy if the lines through it have at least an ϵ fraction of the total weight of all the lines. Then an ϵ-net is defined to be a subset of lines that covers all of the ϵ-heavy points. For any configuration, any ϵ, and any system of weights on the lines through pairs of points in the configuration, one can find an ϵ-net with only $O(\frac{1}{\epsilon}\log\frac{1}{\epsilon})$ lines.[13]

Algorithm 8.20 (Iterated reweighting approximation for LINE-COVER)

If we already know the value of $k = \text{LINE-COVER}(S)$, then a set of $O(k \log k)$ lines that covers all points can be found by the following iterated reweighting algorithm.

1. Assign equal weights to all lines through pairs of points of the given configuration.
2. Repeat:
   a. Let N be a $1/2k$-net of size $O(k \log k)$ for the given weighted lines.
   b. If N covers all points of the configuration, return N as the approximate line cover.
   c. Otherwise, let p be an arbitrary uncovered point.
   d. Double the weights of all the lines through p.

Each repetition of the loop doubles the weight of at least one member of the optimal k-line cover, because at least one line in this optimal cover goes through p. So, on average (even when these weight-doublings are evenly distributed among the members of the optimal cover) the weight of the optimal cover increases by a factor of $2^{1/k}$ per repetition of the loop. On the other hand, each repetition increases the weight of the whole family of lines by a factor less than $(1 + 1/2k)$, because p cannot be $1/2k$-heavy. And

$$2^{1/k} \approx 1 + \frac{\ln 2}{k} \approx 1 + \frac{0.693}{k} > 1 + \frac{1}{2k},$$

so the weight of the optimal cover grows more quickly than the weight of the whole set of lines. Because of these different growth rates, at most $O(k \log n)$ repetitions of the loop are possible. If the algorithm could repeat the loop a larger number of times, then the weight of the optimal cover would become larger than the weight of all of the lines, an impossibility. Therefore, this algorithm eventually finds an approximate line cover.[14]

More typically, k is not already known – it is the number we are trying to approximate, and if we did already know it, why would we need to approximate it? But in this case we can try the same algorithm for successively larger values of k until it succeeds. This produces an approximate line cover with approximation ratio $O(\log k)$, possibly much better than the $O(\log n)$ approximation ratio of the greedy algorithm. This $O(\log k)$ approximation ratio comes from

[13] Vapnik and Chervonenkis (1971); Haussler and Welzl (1987).
[14] Clarkson and Varadarajan (2007).

the $O(\log \frac{1}{\epsilon})$ term in the bound for the size of the ϵ-nets used by this algorithm. Achieving a better than logarithmic approximation is open[15] and could potentially be achieved by solving another open problem on the combinatorial complexity of ϵ-nets for points and lines.[16]

Open Problem 8.21

What is the optimal approximation for LINE-COVER?

Open Problem 8.22

Do points and lines in the plane have ϵ-nets of size $o(\frac{1}{\epsilon} \log \frac{1}{\epsilon})$?

A slightly superlinear lower bound on the size of these ϵ-nets was given by Alon (2012), in a dual setting where the weighted objects are lines and the elements of the net are points. However, Alon's lower bound is still far from the known $O(\frac{1}{\epsilon} \log \frac{1}{\epsilon})$ upper bound.

[15] Dumitrescu and Jiang (2015). [16] Mustafa and Ray (2014).

9 General Position

A set of points is in general position if it forbids the three-point line. Thus, we can define a collection of properties and parameters using this line as an obstacle.

Definition 9.1

We define the property GENERAL-POSITION of being in general position, and parameters MAX-GENERAL for the largest size of a general-position subconfiguration, DELETE-TO-GENERAL for the fewest points to delete to put the remaining points into general position,

$$\text{GENERAL-POSITION} = \text{FORBIDDEN}\big(\text{LINE}(3)\big)$$

$$\text{MAX-GENERAL} = \text{AVOIDS}\big(\text{LINE}(3)\big)$$

$$\text{DELETE-TO-GENERAL} = \text{HITTING}\big(\text{LINE}(3)\big)$$

9.1 No-Three-in-Line

A famous instance of the parameter MAX-GENERAL(S) arises in Dudeney's no-three-in-line problem:

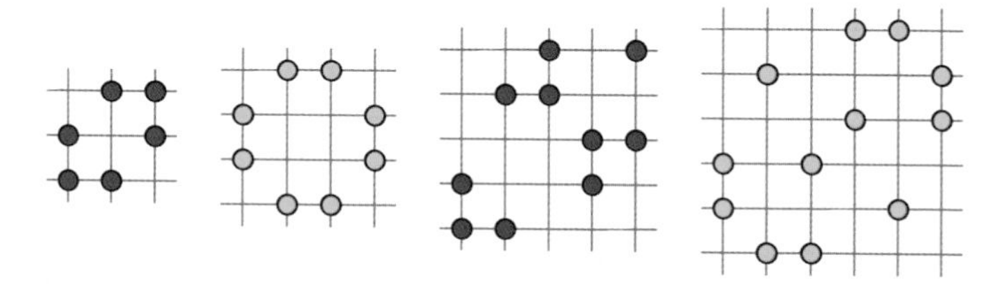

Figure 9.1. Optimal solutions to Dudeney's no-three-in-line problem for $n = 3$, 4, 5, and 6.

Open Problem 9.2 (the no-three-in-line problem)

What is the value of

$$\text{MAX-GENERAL}\big(\text{GRID}(n, n)\big)$$

for each positive integer n?

Dudeney (1917) asked how to place 16 pawns on a chessboard, without allowing any lines of three pawns. It is not just the horizontal, vertical, and diagonal grid lines that count here; every line, of any slope, should have at most two pawns. In our terms, we can think of the chessboard as the configuration GRID$(8, 8)$, and the pawns as a general-position subconfiguration of this grid. To make the problem even harder, Dudeney added that the pawns of the solution should include two at the center of the chessboard, diagonal to each other.

The reason that this puzzle uses 16 pawns is not merely that chess sets include that many. The more relevant fact, in extensions of the puzzle to other grid sizes, is that 16 is two times the width of the chessboard. Doubling the width forms a natural limit to the no-three-in-line problem:

$$\text{MAX-GENERAL}\big(\text{GRID}(n, n)\big) \le 2n.$$

This bound follows from the fact that a general-position subconfiguration of an $n \times n$ grid can only include two points from each grid row. And for small enough values of n, this $2n$ upper bound is matched by solutions that have exactly $2n$ grid points.[1] Figure 9.1 shows some of these solutions. There are many more; the number of different $2n$-point solutions appears to grow exponentially in n. However, it is conjectured that this growth will eventually stop and that, for large n, the largest general-position subconfiguration will have strictly fewer than $2n$ points. More specifically, Guy and Kelly (1968) suggested

[1] Craggs and Hughes-Jones (1976); Kløve (1978, 1979); Anderson (1979); Flammenkamp (1992, 1998).

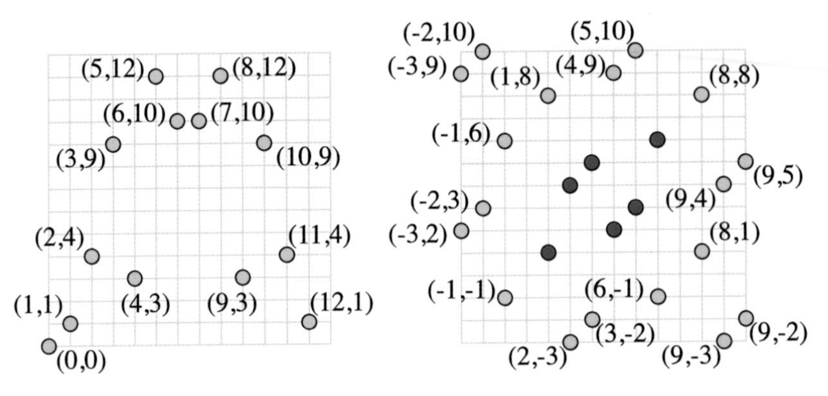

Figure 9.2. The solutions of Erdős (Roth, 1951) (left) and Hall et al. (1975) (right) for the no-three-in-line problem on GRID(14, 14). The left points obey the equation $y = x^2 \bmod 13$. On the right (with a different coordinate system) the yellow solution points obey the equation $xy = 1 \pmod 7$. The red points near the center of the grid obey the same equation, but are omitted by Hall et al. Neither solution is optimal.

that the number of points should be

$$n\sqrt[3]{\frac{2\pi^2}{3}} + o(n) \approx 1.874n + o(n).$$

After Gabor Ellmann pointed out an error in the heuristic reasoning that led to this formula, Guy revised the conjecture[2] to state that the general solution to the no-three-in-line problem should have the form

$$\frac{\pi n}{\sqrt{3}} + o(n) \approx 1.814n + o(n).$$

It is also known that all square grids have general-position subconfigurations of linear size, but with smaller constant factors than 2 or 1.814. The first proof of this was by Erdős,[3] who showed that

$$\text{MAX-GENERAL}\big(\text{GRID}(n, n)\big) \geq n - o(n).$$

Erdős's subconfiguration uses the points with coordinates $(i, i^2 \bmod p)$, for $0 \leq i < p$, where p is the largest prime less than or equal to n. Figure 9.2 (left) shows this construction with $n = 14$ and $p = 13$.

The proof that no three of these points lie on a line involves the arithmetic of mod-p numbers. Erdős's solution points obey a quadratic equation, a mod-p version of the parabola $y = x^2$. And for mod-p numbers, as for the real numbers, Bézout's theorem tells us that a quadratic equation and a linear equation have at most two simultaneous solutions. If Erdős's point set had three points in a Euclidean line, they would also obey a mod-p linear equation, violating Bézout's theorem.

[2] As communicated by Pegg (2005). [3] Roth (1951).

Erdős's construction was improved by Hall et al. (1975), using another quadratic equation modulo a different prime. They select as their prime p the largest prime that is at most $n/2$. They then replace the parabola by a modular hyperbola $xy = k \bmod p$ (for an arbitrary nonzero $k \bmod p$). They divide the grid into 16 smaller squares and choose the points from the hyperbola within 12 of these 16 smaller squares. Figure 9.2 (right) shows the construction for $n = 14$, $p = 7$, and $k = 1$; it generates 18 points compared with Erdős's 13. This construction shows that

$$\textsc{max-general}\big(\textsc{grid}(n, n)\big) \geq \frac{3}{2}n - o(n).$$

Tightening the gap between these constructions and the $2n$ upper bound remains open.

9.2 Finding General-Position Subconfigurations

Testing whether a configuration S is in general position is the same as testing whether $\textsc{online}(S) \geq 3$, which can be solved in time $O(n^2)$ as discussed in Section 8.4. However, finding large general-position subconfigurations (or small sets of points to delete to make the remaining points be in general position) is harder.

Theorem 9.3

Computing MAX-GENERAL *or* DELETE-TO-GENERAL *is* NP*-hard and approximating either is* APX*-hard.*

Proof. We defer the proof to Section 15.3. □

We will return to the approximation of these two parameters in Section 9.5. From the general form of their definition in terms of AVOIDS and HITTING, parameters MAX-GENERAL and DELETE-TO-GENERAL both have polynomial obstacle size (Observation 7.5 and Theorem 7.9). For small parameter values, we can be more specific.

Example 9.4

The obstacles for $\textsc{delete-to-general}(S) < 2$ are configurations that cannot be made general-position by removing only one point by itself, because otherwise they would not be obstacles. But, for every one point p that might be removed, there exists another point q such that removing both p and q

leaves a general position subconfiguration, because otherwise p could be removed to get a smaller obstacle. These obstacles are

- LINE(4)
- the four configurations consisting of two disjoint lines of three points (Figure 8.7), and
- the 16 six-point configurations that have three or four lines of three points (Figure 8.6).

By Theorem 7.9, DELETE-TO-GENERAL is fixed-parameter tractable. The same is true for MAX-GENERAL.

Theorem 9.5

Computing MAX-GENERAL(S) *is fixed-parameter tractable.*

Proof. The following case analysis can be used as the basis for a kernelization algorithm. Suppose we are trying to test whether MAX-GENERAL(S) $< k$. Let L be any line through three or more points of S. Then any general-position subconfiguration of S can contain at most two points of L. If P is an arbitrary partial solution to the problem (any general-position subconfiguration of S with $|P| < k$), then one of the following two cases occurs with respect to L:

- P may already contain two points of L. In this case no more points of L can be added to P while preserving its general position.
- P may contain zero or one points of L and could potentially be augmented by adding more points from L. In this case, because $|P| < k$, the lines through pairs of points in P cross L at most $\binom{k-1}{2}$ times. And because P contains at most one point from L, none of these lines coincides with L. Therefore, if we wish to augment P by one or two points of L, we may choose these added points to be any of the points that is not located at one of these $\binom{k-1}{2}$ crossings.

Based on this analysis, we may test whether MAX-GENERAL(S) $< k$ or whether MAX-GENERAL(S) $\geq k$ using an algorithm that performs the following steps.

Algorithm 9.6 (Kernelization for MAX-GENERAL)

1. Let G be an empty set of points, and for each point p of S, add p to G if the resulting set would remain in general position. The resulting set G is a

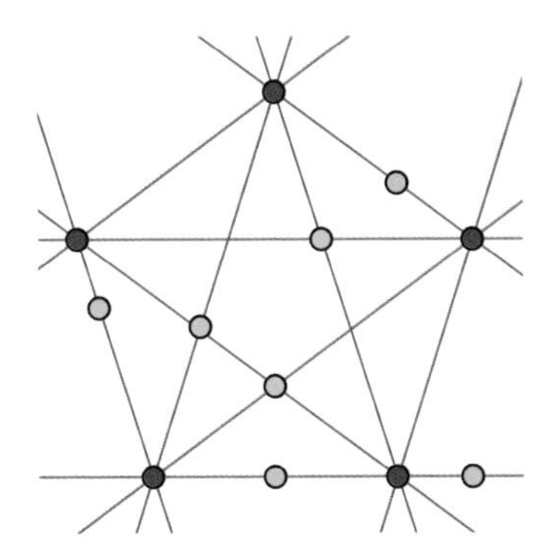

Figure 9.3. A subconfiguration (red) is maximal for the property of being in general position (meaning that no more points can be included in it while preserving its general position) if and only if all other points lie on lines through pairs of points in the subconfiguration.

maximal general-position subconfiguration G of S. That is, G is a subconfiguration of S that is in general position and is not a subset of any larger general-position subconfiguration of S.

2. If $|G| \geq k$, we have found that MAX-GENERAL$(S) \geq k$ and solved the problem directly, so we may stop the algorithm.
3. In the remaining case, S may be covered by the $O(k^2)$ lines through pairs of points in G (Figure 9.3). For otherwise G would not be maximal: any uncovered point could have been added to G. Let C be this set of covering lines.
4. While C includes a line that contains at least $\binom{k-1}{2} + 2$ points, choose L (arbitrarily) as one of these lines. Remove all points of L from the given input configuration S and reduce the parameter k by one. This transformation is safe to do, as any optimal solution to the resulting reduced problem may be augmented, by adding back one or two points from L, to become an optimal solution to the original problem.
5. The loop in the previous step terminates when it can no longer find a line L in C that contains many points. When this happens, let K be the remaining configuration. Then K is covered by $O(k^2)$ lines in C, each of which contains $O(k^2)$ points of K. Therefore, K has size $O(k^4)$.
6. For each line L that was removed in the earlier loop, in the reverse order of the removal, add one to k and then modify K by adding to it exactly $\binom{k-1}{2} + 2$ points of L, chosen arbitrarily. This number of points is enough to ensure that an optimal solution to K can always be extended to an optimal solution of $K \cup L$.
7. Use a brute force search to find the value of MAX-GENERAL for the kernel K.

The kernel K is formed from $O(k^4)$ points by adding $O(k)$ subsets of $O(k^2)$ points, so it has total size $O(k^4)$. Other than the final brute force search, all steps take polynomial time. Therefore, this is a valid kernelization that shows the problem to be fixed-parameter tractable. □

Theorem 9.7

The configurations S for which MAX-GENERAL$(S) < 5$ *are well-quasi-ordered, and the number of such configurations with n points is polynomial in n. Any monotone parameter of these configurations is nonuniform fixed-parameter tractable. However, the configurations for which* MAX-GENERAL$(S) < 7$ *are not well-quasi-ordered, and the number of such configurations is exponential in $n \log n$.*

Proof. For the results on configurations with MAX-GENERAL$(S) < 5$, see Section 17.6. For the results on configurations with MAX-GENERAL$(S) < 7$, see Section 15.4. □

Open Problem 9.8

Are the configurations with MAX-GENERAL$(S) < 6$ well-quasi-ordered, and how many such configurations are there?

9.3 Property Testing

The property testing algorithm of Theorem 6.8 gives a tight bound on the sample size when used to test for being in general position.

Theorem 9.9

Sample-based property testing for being in general position requires samples of size $\Theta(n^{2/3})$ to achieve a constant false positive rate for configurations that are a constant distance ϵ from being in general position.

Proof. Theorem 6.8 shows that samples of this size are sufficient. To show that samples of this size are necessary, consider an input configuration S consisting of $n/3$ triples of collinear points, with no other collinearities than the ones in those triples. (Such a configuration is illustrated in Figure 11.7.) Then S is $1/3$-far from being in general position. Sample-based property testing on S will generate a false positive unless the sample includes all three points of at least one triple. For samples of size $cn^{2/3}$, the probability that it includes the points of any one triple is c^3/n, so the expected number of triples of this form that it includes is $c^3/3$. If this expected number is less than $1 - p$, then by Markov's inequality[4] the algorithm will have a false positive rate greater than p. So, c needs to be at least a constant for the algorithm to achieve the required false positive rate. □

9.4 Inequalities

As the following observation shows, every algorithm that is fixed-parameter tractable with respect to MAX-GENERAL is also fixed-parameter tractable with respect to LINE-COVER.

Observation 9.10

MAX-GENERAL $\doteq$ LINE-COVER.

Proof. Recall that this notation means that the two parameters are bounded above and below by functions of each other. In this case, they obey the inequalities

$$\text{MAX-GENERAL}(S)/2 \leq \text{LINE-COVER}(S) \leq \binom{\text{MAX-GENERAL}(S)}{2},$$

for every configuration S. The first inequality follows because, in any general-position subconfiguration of S, each line can only cover two points. The second follows because, for any maximum (or maximal) general-position subconfiguration of S, the remaining points must all lie on the lines through pairs of points from the subconfiguration (Figure 9.3). □

Observation 9.11

ONLINE $\ll$ DELETE-TO-GENERAL.

Proof. More specifically, for every configuration S,

$$\text{ONLINE}(S) \leq \text{DELETE-TO-GENERAL}(S) + 2,$$

[4] For a nonnegative integer random variable, the probability of being nonzero is at most its expected value.

because all but two points of the heaviest line of S must be removed to achieve general position. To construct a family of configurations S with $\textsc{online}(S)$ bounded and $\textsc{delete-to-general}(S)$ unbounded, consider the disjoint union of $n/3$ three-point lines, in general position with respect to each other. (Figure 11.7 depicts a configuration of this form, which we already used in Theorem 9.9 to prove a lower bound for property testing.) □

Erdős (1986, 1988) asked for a trade-off between ONLINE and MAX-GENERAL that is valid for all configurations S and as tight as possible. The square grid shows that there exists a configuration S in which both of these parameters are $O(\sqrt{|S|}\,)$. For any maximal general-position subconfiguration of any configuration S, the other points of S lie on one of the lines determined by the subconfiguration (Figure 9.3 again), so for all configurations S,

$$|S| \le \textsc{max-general}(S) + \binom{\textsc{max-general}(S)}{2}(\textsc{online}(S) - 2).$$

This shows that the larger of ONLINE and MAX-GENERAL must be proportional to the cube root of $|S|$. This reasoning proves the following observation.

Observation 9.12

$\textsc{size} \doteq \textsc{online} + \textsc{max-general}.$

However, there remains a gap between the cube-root lower bound on the larger of ONLINE and MAX-GENERAL demonstrated above, and the square-root upper bound given by the square grids. We would like to close this gap. After earlier advances on the solution to this problem by Füredi (1991) and Lefmann (2012), Payne and Wood (2013) found tighter relations. Their results show that (as in the grid) the larger of the two quantities ONLINE and MAX-GENERAL is always nearly proportional to the square root of the size. More precisely, they proved the following.

Theorem 9.13 (Payne and Wood, 2013)

Let S be an arbitrary configuration. If $\textsc{online}(S) = O(\sqrt{|S|})$, *then*

$$\textsc{max-general}(S) = \Omega\left(\sqrt{\frac{|S|}{\log \textsc{online}(S)}}\right).$$

If (for any constant $\epsilon > 0$*)* $\textsc{online}(S) = O(|S|^{1/2-\epsilon})$, *then*

$$\textsc{max-general}(S) = \Omega\left(\sqrt{\frac{|S| \log |S|}{\log \textsc{online}(S)}}\right).$$

If ONLINE$(S) = \Omega(\sqrt{|S|})$, *then*

$$\text{MAX-GENERAL}(S) = \Omega\left(\frac{|S|}{\text{ONLINE}(S)}\right).$$

The proof of Theorem 9.13 involves the fact that when ONLINE is sufficiently small, there can be only a near-quadratic number of collinear triples of points, The proof begins by reducing the problem to one in which no point belongs to significantly more collinear triples than the average. Then, it uses a powerful tool from probabilistic combinatorics, the *Lovász local lemma*,[5] to partition the points into a small number of general-position subconfigurations, the largest of which has at least the stated number of points. For an algorithmic version of this proof, see Section 9.6.

9.5 Approximation

In this section we describe approximation algorithms below for the parameters of this chapter. They are weak, too weak to help solve the no-three-in-line problem. But they are the best we know, and at least they achieve a nontrivial approximation ratio.

Algorithm 9.14 (Greedy approximation for MAX-GENERAL**)**

To approximate MAX-GENERAL, we greedily find a maximal general-position subconfiguration. That is, we perform the following steps.

1. Initialize G to be the empty configuration.
2. For each point p of S, if adding p to G would keep G in general position, then add it.
3. Return G as the approximate maximum general-position subconfiguration, and $|G|$ as the approximation to MAX-GENERAL(S).

Lemma 9.15

The greedy approximation algorithm for MAX-GENERAL *constructs a general-position subconfiguration G with size*

$$|G| \geq \sqrt{\max\left\{\text{MAX-GENERAL}(S), \frac{2|S|}{\text{ONLINE}(S)}\right\}}.$$

[5] Erdős and Lovász (1975); Spencer (1977); Shearer (1985).

Proof. Each point of S is covered by at least one of the $\binom{|G|}{2}$ lines through pairs of points in G. Each such line contains at most two points of the maximum general-position subconfiguration, and at most ONLINE(S) points of S. Therefore,

$$\frac{|G|(|G|-1)}{2} = \binom{|G|}{2} \geq \max\left\{\frac{\text{MAX-GENERAL}(S)}{2}, \frac{|S|}{\text{ONLINE}(S)}\right\}.$$

The lemma follows by multiplying both sides by two and taking square roots. □

Theorem 9.16

We can approximate MAX-GENERAL(S), *for a configuration* S, *to within an approximation ratio of* $O(\sqrt{|S|})$ *in polynomial time.*

Proof. We use the greedy approximation algorithm for MAX-GENERAL described above. Because its size is at least $\sqrt{\text{MAX-GENERAL}(S)}$, its approximation ratio is at most $\sqrt{\text{MAX-GENERAL}(S)} \leq \sqrt{|S|}$. □

Algorithm 9.17 (Greedy approximation for DELETE-TO-GENERAL)

We can also use another greedy algorithm to obtain a much more accurate approximation to DELETE-TO-GENERAL. It performs the following steps.

1. Initialize R (the set of points to remove from S) to be empty.
2. While $S \setminus R$ is not in general position:
   a. Find a triple of collinear points (p, q, r) in $S \setminus R$.
   b. Add all three points to R.
3. Return R as the approximate set of points to be removed to make the remaining points of S be in general position, and $|R|$ as our approximation to DELETE-TO-GENERAL(S).

Theorem 9.18

We can approximate DELETE-TO-GENERAL(S), *for a configuration* S, *to within an approximation ratio of* 3 *in polynomial time.*

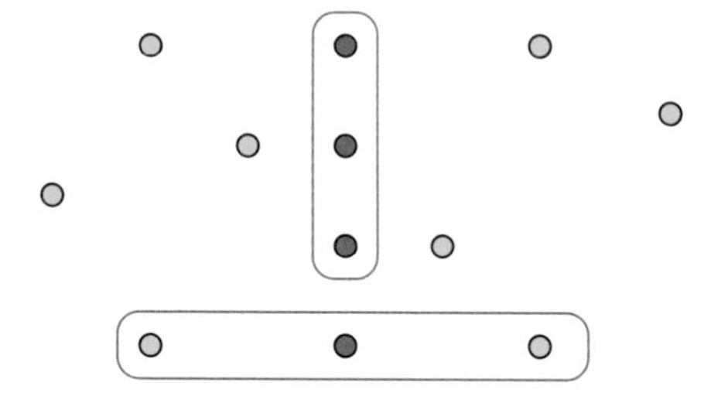

Figure 9.4. The greedy approximation to DELETE-TO-GENERAL of Theorem 9.18 chooses triples of collinear points (red ovals) and removes all three points in each oval, until no more triples can be found. In this example, it removes six points. The optimal solution removes the four blue points instead. Each oval chosen in the approximation contains at least one blue point, so the approximate solution is within a factor of three of optimal.

Proof. For each triple that the greedy algorithm adds to its approximate solution, at least one point of the triple belongs to the optimal solution (Figure 9.4). Therefore, R is within a factor of three of optimal. □

The same approach of repeatedly finding a forbidden subconfiguration and removing all of its points will approximate any parameter HITTING$(C_1, C_2, \ldots)$ to within a factor of $\max(|C_1|, |C_2|, \ldots)$. This algorithm achieves an approximation ratio of three for DELETE-TO-GENERAL, because the size of the forbidden configuration LINE(3) used to define DELETE-TO-GENERAL is three.

Open Problem 9.19

What approximation ratio for MAX-GENERAL(S) can a polynomial-time approximation algorithm achieve?

9.6 Entropy Compression

In Theorem 9.13 we saw that in any configuration S, as long as ONLINE(S) is smaller than $\sqrt{|S|}$, then S will contain a general-position subconfiguration whose size is large, near $\sqrt{|S|}$. But the greedy algorithm of Theorem 9.16 is not as good: it produces general-position subconfigurations of that size only when ONLINE(S) is constant or near-constant. In this section we describe a more complex randomized algorithm that produces larger general-position subconfigurations. It does so by adapting the (nonconstructive) proof of Theorem 9.13 by Payne and Wood (2013), replacing its use of the Lovász local lemma by a related technique called *entropy compression*.[6]

[6] Moser and Tardos (2010).

Lemma 9.20 (Payne and Wood, 2013)

For any configuration S, the number of collinear triples of points in S is

$$O\left(|S|^2 \log \text{ONLINE}(S) + |S| \, \text{ONLINE}(S)^2\right).$$

The proof of Lemma 9.20 invokes the Szemerédi–Trotter theorem, according to which n points and m lines can together have at most $O(n^{2/3}m^{2/3} + n + m)$ point–line incidences.[7] This limits the number of distinct i-point lines, for each i up to $\text{ONLINE}(S)$, from which the number of collinear triples coming from each such line can also be bounded.

Corollary 9.21

In polynomial time, for any configuration S, we can find a subconfiguration S′ with size |S|/2 such that each point of S′ belongs to

$$O\left(|S| \log \text{ONLINE}(S) + \text{ONLINE}(S)^2\right)$$

collinear triples.

Proof. Compute the number of collinear triples containing each point, and select a subconfiguration of $|S|/2$ points for which this number is at most the median. If the number of triples per point exceeded the bound of the corollary, the remaining (unselected) points would belong to so many triples that their total would exceed the bound of Lemma 9.20. □

Algorithm 9.22 (Entropy compression for finding large general-position subconfigurations)

Our algorithm chooses a coloring of the configuration S' resulting from Corollary 9.21, and then repeatedly uses a recursive subroutine $\text{FIX}(t)$ to randomly recolor any collinear triple t of points that all have the same color. When this subroutine terminates, t and all other triples whose coloring was modified will have at least two distinct colors. If all triples can be recolored in this way, the color classes will form a partition of S' into general-position subconfigurations, the largest of which will match the bound of Theorem 9.13.

[7] Szemerédi and Trotter (1983); Székely (1997).

In more detail, we perform the following steps.

1. Use Corollary 9.21 to find a large subconfiguration S' within which each point belongs to $O(|S| \log \text{ONLINE}(S) + \text{ONLINE}(S)^2)$ collinear triples.
2. Let s be the maximum number of collinear triples containing a single point of S', and choose $k = \lceil c\sqrt{s} \rceil$, for a sufficiently large constant c to be determined later.
3. Choose an arbitrary k-coloring of S'.
4. For each collinear triple t in S', if t has only one color, call $\text{FIX}(t)$.
5. Count the number of points with each color and return the largest color class.

The recursive subroutine $\text{FIX}(t)$ performs the following steps.

1. Choose a random k-coloring of the three points in t.
2. For each collinear triple t' that overlaps t and has points of only one color (possibly including t itself, after the first step), call $\text{FIX}(t')$ recursively.

If FIX always terminates in polynomial time, then so does the whole algorithm, returning a general-position subconfiguration of size $|S|/2k$. It remains to show that, for sufficiently large values of the constant c, we can guarantee that FIX does terminate in polynomial time, rather than taking longer or even going into an infinite recursion.

To show this, we observe that, if a recursive call to FIX recolors r triples of points, then it generates $3r \log k$ bits of entropy in the random choices it makes. We can recover all of this information by recording the sequence of coloring steps of the algorithm in a different way. For each call to $\text{FIX}(t)$, record the color that the three points of t had at the start of the call ($\log k$ bits of information), and the sequence of overlapping triples that are called recursively from this call ($\log 3s$ bits of information per recursive call, based on the fact that each must be one of the triples involving a point of t). Additionally, record the argument to the top-level call to FIX ($3 \log |S|$ bits of information to list the three points of this triple) and the final coloring after the r triples have been recolored ($|S| \log k$ bits of information). This information can be played backward from the final coloring to reconstruct all of the random choices made by the algorithm, from which it follows that

$$r(\log s + \log k + \log 3) + 3 \log |S| + |S| \log k \geq 3r \log k.$$

But $\log s \leq 2(\log k - \log c)$. Using this inequality to substitute $\log k$ for $\log s$ shows that the multiplier of r on the left is at most $3 \log k + \log 3 - 2 \log c$, while the multiplier on the right is just $3 \log k$. If we choose $c > \sqrt{3}$, the left multiplier will be smaller than the right multiplier, and we will save a constant number of bits of information on the left hand side of the inequality, compared with the right hand side, per recursive call. This can only happen $O(|S| \log k)$ times before the left-hand side becomes smaller than the right-hand side, so to prevent a contradiction FIX must terminate after $O(|S| \log k)$ recursive calls.

This argument completes the proof of the main result of this section.

Theorem 9.23

In randomized polynomial time, for any configuration S, we can find a general-position subconfiguration of S whose size is

$$O\left(\min\left(\sqrt{\frac{|S|}{\log \textsc{online}(S)}}, \frac{|S|}{\textsc{online}(S)}\right)\right).$$

10 General-Position Partitions

As with the parameters defined in the previous chapter, we may use the obstacle LINE(3) to define a parameter GENERAL-PARTITION that counts the fewest subconfigurations in a partition into general-position sets:

$$\text{GENERAL-PARTITION} = \text{PARTITION}\big(\text{LINE}(3)\big).$$

The values of this parameter on grids were studied by Wood (2004) as a "natural generalisation" of the no-three-in-line problem. By the known bounds for the no-three-in-line problem, each subconfiguration in a general-position partition of GRID(n, n) has at most $2n$ points, so to cover all n^2 points of the grid, at least $\lceil n/2 \rceil$ subconfigurations are needed. Wood used a decomposition into translated copies of Erdős's construction for the no-three-in-line problem to show that this number is at most $2n + o(n)$. He asks the following question.

Open Problem 10.1 (Wood, 2004)

What is the largest constant c such that

$$\text{GENERAL-PARTITION}\big(\text{GRID}(n, n)\big) \le cn + o(n)?$$

10.1 Glimpses of Higher Dimensions

Although higher dimensional grids are not planar point sets, they can be made planar by a linear projection. And if we choose the linear projection carefully, we can avoid creating any new lines of points, beyond the ones that already existed in the high-dimensional point set. Figure 10.1 depicts a

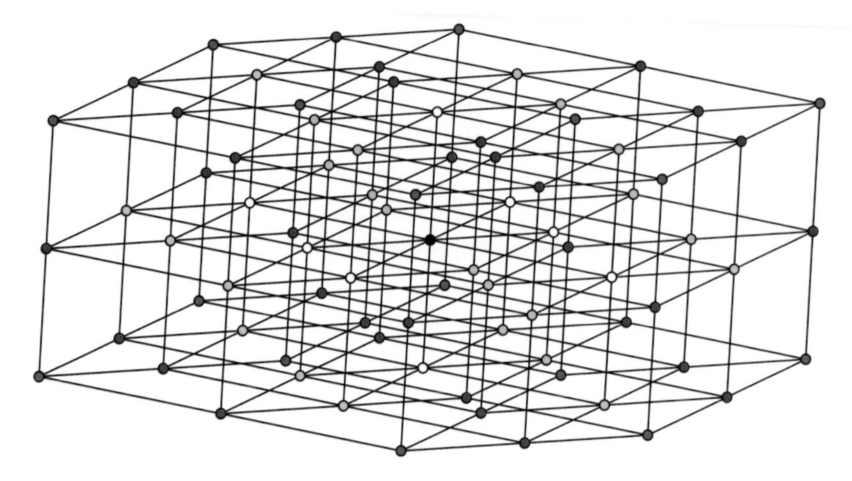

Figure 10.1. A linear projection of $\{1, 2, 3\}^4$ onto the plane. The proof of Observation 10.11 uses configurations constructed in this way from higher-dimensional projections. The coloring shows a partition into five general-position subconfigurations, obtained by partitioning the points according to how many of their original coordinates equal 2.

four-dimensional example, in which we have projected a $3 \times 3 \times 3 \times 3$ grid into the plane.

Definition 10.2

For any positive integer d, we define $\text{TERNARY}(d)$ to be the class of configurations formed by linearly projecting the set of points $\{1, 2, 3\}^d$ from d-dimensional Euclidean space to the plane, without creating any new collinear triples of points that were not already collinear in d-dimensional space.[1]

Many different configurations may be generated in this way, but the distinctions between them do not matter for the purposes of examining general-position subconfigurations. From now on, we will treat $\text{TERNARY}(d)$ as if it is a single configuration rather than a class of configurations. Although the lines in these configurations can be hard to visualize (we have shown only some of them in the figure), it is easy to describe them symbolically.

Observation 10.3

In $\text{TERNARY}(d)$, *each three-point line can be described in exactly two ways by a string of length d over the alphabet* $\{1, 2, 3, \uparrow, \downarrow\}$ *that includes at least one*

[1] This construction appears in Füredi (1991).

arrow symbol. Given a string of this type, we may find the ith point on its line (for $i = 1, 2,$ or 3) by replacing each up arrow by i, replacing each down arrow by $4 - i$, and projecting the resulting d-tuple of numbers to the plane. The two strings that represent each line can be formed from each other by reversing the directions of all their arrows.

TERNARY(d) has exactly $(5^d - 3^d)/2$ three-point lines, because there are 5^d strings of these five symbols, 3^d of which do not have any arrows, and each line is represented by two different strings. For example, the standard tic-tac-toe board $\{1, 2, 3\}^2$ has three horizontal lines $\uparrow 1$, $\uparrow 2$, and $\uparrow 3$, three vertical lines $1\uparrow$, $2\uparrow$, and $3\uparrow$, and two diagonal lines $\uparrow\uparrow$ and $\uparrow\downarrow$, for a total of $8 = (5^2 - 3^2)/2$ lines. The points of the line $\uparrow\downarrow$ are $(1, 3)$, $(2, 2)$, and $(3, 1)$, and the same line may also be described by reversing the arrows (and the ordering of its points) as the string $\downarrow\uparrow$. No lines are longer than three points; this can be stated more formally in the following observation.

Observation 10.4

For all d, ONLINE(TERNARY(d)) $= 3$.

The two-dimensional case, GRID$(3, 3)$, is familiar as the starting set of positions in the game of tic-tac-toe, and its three-point lines are the lines used to win in tic-tac-toe. Unfortunately, the higher-dimensional TERNARY configurations do not make good tic-tac-toe boards. Even the three-dimensional board TERNARY(3) is too easy for the first player to win. This follows from the following observation.

Observation 10.5

For all $d \geq 3$,

$$\text{GENERAL-PARTITION}(\text{TERNARY}(d)) > 2.$$

Proof. For $d = 3$, see Beck (2008, exercise 3.3, p. 51). For larger d, this follows from the fact that TERNARY(3) is a subconfiguration of TERNARY(d). □

If we use a high-dimensional grid or its projection TERNARY(d) as a tic-tac-toe board, someone must win: by Observation 10.5, it is impossible for the game to end in a tie, because a drawn position in tic-tac-toe can only result when each player has marked a collection of cells in general position. Motivated by this application of general-position partitions to high-dimensional tic-tac-toe,

Hales and Jewett (1963) proved a stronger form of Observation 10.5 that we will use in Observation 10.11 to compare GENERAL-PARTITION with ONLINE. (Their theorem applies to grids of any side length, but we will only use the version for ternary grids.)

Theorem 10.6 (The Hales–Jewett theorem (Hales and Jewett, 1963))

For every k there exists a d such that

$$\text{GENERAL-PARTITION}\big(\text{TERNARY}(d)\big) > k.$$

More strongly, Hales and Jewett showed that, for large enough d, every k-partition of TERNARY(d) will include a line described by a string of the form $\{\uparrow, 1, 2, 3\}^d$, without down arrows. Their theorem implies that in high-dimensional tic-tac-toe, even when played by k different players who are all trying to reach a drawn game, one of the players will win.

The dimension d that can be shown to cause a line to exist (as a function of k, the number of sets in the partition) is enormous. It is much more quickly growing than exponential, or even than a tower of powers of two, k levels high. It belongs to the class of so-called primitive recursive functions, at the fifth level of the Grzegorczyk hierarchy. The first level of this hierarchy includes multiplication by constants; the second level includes the polynomial functions; and the third level includes exponentiation to any constant number of levels, such as the function that maps n to $2^{2^{2^n}}$. Variable-height towers of powers are at the fourth level, and the best known bound for the Hales–Jewett theorem is at most at the fifth level.[2] This upper bound on d is far from the following, the best lower bound that we know of.

Observation 10.7

For all k, and all $d < k$,

$$\text{GENERAL-PARTITION}\big(\text{TERNARY}(d)\big) \le k.$$

Proof. Partition the points according to how many of their coordinates (prior to the projection) are equal to 2. There are $d+1$ choices for this number, so there are $d+1$ sets in the partition. In any three-point line of $\{1, 2, 3\}^d$, the middle point has more coordinates equal to 2 than the other two points, so this partition breaks up every line. □

[2] Shelah (1988).

The gap between these bounds raises the following question.

Open Problem 10.8

How big is GENERAL-PARTITION(TERNARY(d))?

We can also ask about MAX-GENERAL for the TERNARY configurations. By taking the largest subconfiguration in the partition of Observation 10.7 (the one in which roughly 1/3 of the coordinates are twos), we can show the following.

Observation 10.9

$$\text{MAX-GENERAL}\big(\text{TERNARY}(d)\big) \geq \binom{d}{\lfloor d/3 \rfloor} 2^{\lceil 2d/3 \rceil} = \Omega\left(\frac{3^d}{\sqrt{d}}\right).$$

In the other direction, Füredi (1991) observed that a "density version" of the Hales–Jewett theorem[3] can be used to show that

$$\text{MAX-GENERAL}\big(\text{TERNARY}(d)\big) = o(3^d).$$

Open Problem 10.10

Is Observation 10.9 tight, or does TERNARY(d) have even larger general-position subconfigurations?

These questions on the size of the general-position subconfigurations of TERNARY(d) are closely related to another topic of current interest in discrete geometry, the problem of *cap-sets*. A cap-set is a general-position subconfiguration of a space $\mathbb{Z}_3^d$ that resembles TERNARY(d) but has more lines. In $\mathbb{Z}_3^d$, three points form a line when, in each coordinate position, either all three points have equal coordinates or all three have distinct coordinates.

This space $\mathbb{Z}_3^d$ and its cap-sets have become familiar through the card game Set, whose 81 cards each have three possible values of their suit (shape of the symbols on the card), number (of symbols shown on the card), color, and texture (outlined, shaded, or solid).[4] The goal of this game is to collect lines of three

[3] Proved by Furstenberg and Katznelson (1989). [4] McMahon et al. (2016).

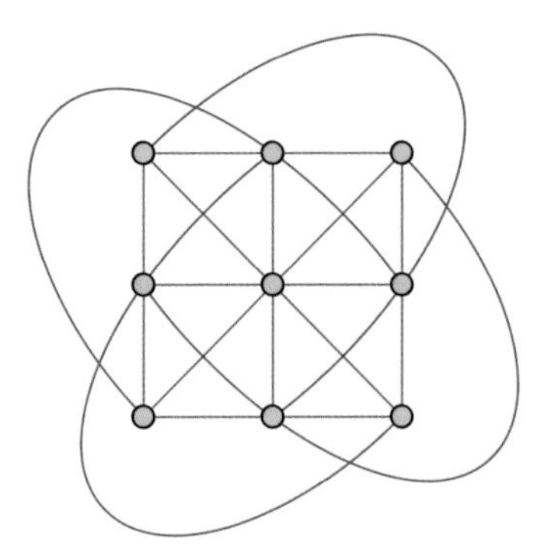

Figure 10.2. The Hesse configuration $\mathbb{Z}_3^2$ adds four more three-point lines (shown as red curves) to GRID(3, 3). It cannot be realized by points in the Euclidean plane, because it has no ordinary (two-point) lines.

cards, as defined in $\mathbb{Z}_3^4$ using these attributes as coordinates. A cap-set describes a collection of cards from which no line can be collected.

Unfortunately, the extra lines of $\mathbb{Z}_3^d$ prevent it from being realized (with the same collinearities) by points in the plane. Even $\mathbb{Z}_3^2$ (the *Hesse configuration*,[5] Figure 10.2) has no realization. This unrealizability happens because every two points of $\mathbb{Z}_3^d$ are part of a three-point line. A realization of this space (for $d \geq 2$) would violate the Sylvester–Gallai theorem, according to which every noncollinear planar configuration must include at least one two-point line.

Cap-sets in $\mathbb{Z}_3^d$ must be significantly smaller than the large general-position subconfigurations of TERNARY(d) that we describe in Observation 10.9. The largest cap-sets have size exponential in d, but the base of the exponent is strictly between 2 and 3.[6]

10.2 Inequalities

Observation 10.11

ONLINE $\ll$ GENERAL-PARTITION.

Proof. The inequality

$$\text{ONLINE}(S) \leq 2 \cdot \text{GENERAL-PARTITION}(S)$$

(for all configurations S) follows from the fact that, if L is any line that contains ONLINE(S) points of S, then at most two points from each general-position subconfiguration can belong to L. Therefore, just to cover the points

[5] Hesse (1844). [6] Edel (2004); Croot et al. (2016); Ellenberg and Gijswijt (2017).

of L, the optimal partition into general-position subconfigurations needs $\text{ONLINE}(S)/2$ subconfigurations. This inequality implies that either $\text{ONLINE} \ll \text{GENERAL-PARTITION}$ or $\text{ONLINE} \doteq \text{GENERAL-PARTITION}$. To see that the first of these two possibilities is the correct one, consider the configurations $\text{TERNARY}(i)$ studied in Section 10.1, which have bounded values of $\text{ONLINE}(S)$ but unbounded values of $\text{GENERAL-PARTITION}(S)$. □

A stronger inequality between ONLINE and GENERAL-PARTITION follows from a recent result of Balogh and Solymosi.

Theorem 10.12 (Balogh and Solymosi, 2017)

For every constant $\epsilon > 0$, *there exist arbitrarily large configurations* S *with* $\text{ONLINE}(S) = 3$ *and*

$$\text{MAX-GENERAL}(S) = O\big(|S|^{5/6+\epsilon}\big).$$

Corollary 10.13

For every constant $\epsilon > 0$, *there exist arbitrarily large configurations* S *with* $\text{ONLINE}(S) = 3$ *and*

$$\text{GENERAL-PARTITION}(S) = \Omega\big(|S|^{1/6-\epsilon}\big).$$

There still remains a large gap between the exponents of $1/6$ and $5/6$ given by these results for GENERAL-PARTITION and MAX-GENERAL (respectively), and the exponents of $1/2$ given by Theorem 9.13.

Open Problem 10.14

For configurations S that have $\text{ONLINE}(S) = 3$, what is the largest possible value of $\text{GENERAL-PARTITION}(S)$ and the smallest possible value of $\text{MAX-GENERAL}(S)$, as a function of $|S|$?

10.3 Complexity

Computing GENERAL-PARTITION is NP-hard, even for very small parameter values. As a step toward proving this, we describe the behavior of GENERAL-PARTITION on a special configuration shown in Figure 10.3.

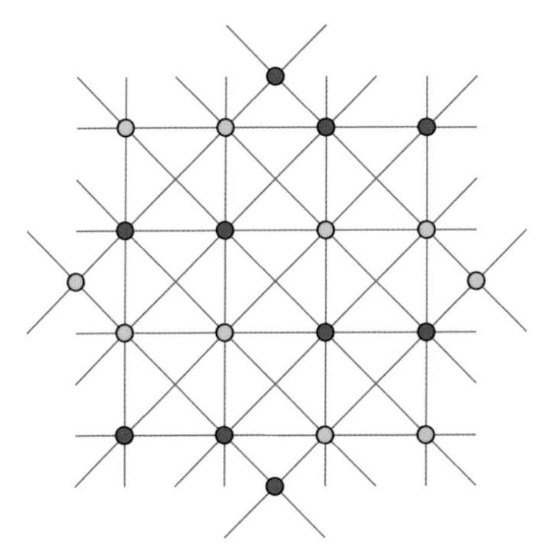

Figure 10.3. A configuration partitioned into two general-position subconfigurations (GENERAL-PARTITION$(S) = 2$). The top left, top right, bottom left, and bottom right points of the central 4×4 grid alternate between the two subconfigurations. Lemma 10.15 shows that this is true for all such partitions of this configuration.

Lemma 10.15

In every partition of the configuration in Figure 10.3 into two general-position subconfigurations, the four points that form the vertices of its outermost axis-aligned square must alternate between the two subconfigurations, with one diagonal pair in one subconfiguration and the other diagonal pair in the other subconfiguration.

Proof. Suppose for a contradiction that this configuration is partitioned into two general-position subconfigurations without these four points alternating in this way between the two subconfigurations, and rotate the configuration so that two nonalternating vertices of the outer square are on the top side of the square. Number the two subconfigurations of the partition as 0 and 1, with the two nonalternating vertices as being in subconfiguration 0, so that the topmost horizontal line of the configuration, through four points, is numbered 0110. We consider the following cases.

- If the second-to-top horizontal line is numbered $01{*}{*}$ (where the $*$ symbols label positions that are irrelevant for this case), then the two zeros on the leftmost vertical line of the configuration force the third-from-top horizontal line to be numbered $1{*}{*}{*}$. But this produces a diagonal line of 1 positions, violating the assumption that subconfiguration 1 is in general position. The same reasoning applies by symmetry when the second-to-top line is numbered ${*}{*}10$.

- If the second-to-top horizontal line is numbered $*01*$, then the vertical line with two ones forces the third-from-top horizontal line to be numbered $**0*$. But this produces a diagonal line of 0 positions and another contradiction. The same reasoning applies by symmetry when the second-to-top line is numbered $*10*$.
- The only remaining way to number the second-to-top horizontal line with two zeros and two ones (so it is itself partitioned into general position subconfigurations) is to use the numbering 1001. Then the two main diagonals of the GRID(4) subconfiguration have two zeros on them, forcing the remaining numbers on those diagonals to be ones. Filling in the remaining numbers on the grid so that each vertical line is partitioned into two zeros and two ones, the numbering for the whole grid is forced to be the matrix

$$\begin{bmatrix} 0 & 1 & 1 & 0 \\ 1 & 0 & 0 & 1 \\ 0 & 1 & 1 & 0 \\ 1 & 0 & 0 & 1 \end{bmatrix}.$$

  However, there is no way to consistently assign the remaining four off-grid points to the two subconfigurations.

Thus, all cases end in a contradiction to the assumption that the outer four points do not alternate as the lemma says they do. □

Theorem 10.16

The property of having GENERAL-PARTITION < 3 *has infinitely many obstacles and is* NP*-complete to test.*

Proof. We use configurations built from multiple copies of the configuration in Figure 10.3. By positioning one of these copies so that two previously placed points are vertices of its outer axis-aligned square, we can use Lemma 10.15 to infer that these two points must be on the same side of any partition into two subconfigurations, or on different sides, depending on how we place the copy.

To find infinitely many obstacles, use copies of this subconfiguration to connect a set of points into a cycle, with an odd number of cycle edges replaced by different-side connections and the remaining edges replaced by same-side connections. See Figure 10.4 for an example. If necessary, remove points from each copy to make it be minimal with respect to its forcing properties. If the initial cycle's vertices are placed in sufficiently general position, there will be no extra collinearities beyond the ones in each copy.

To prove NP-completeness, we describe a reduction (that is, a polynomial-time transformation from an input to one problem into an input to another problem, preserving the correct output) that transforms a known NP-complete

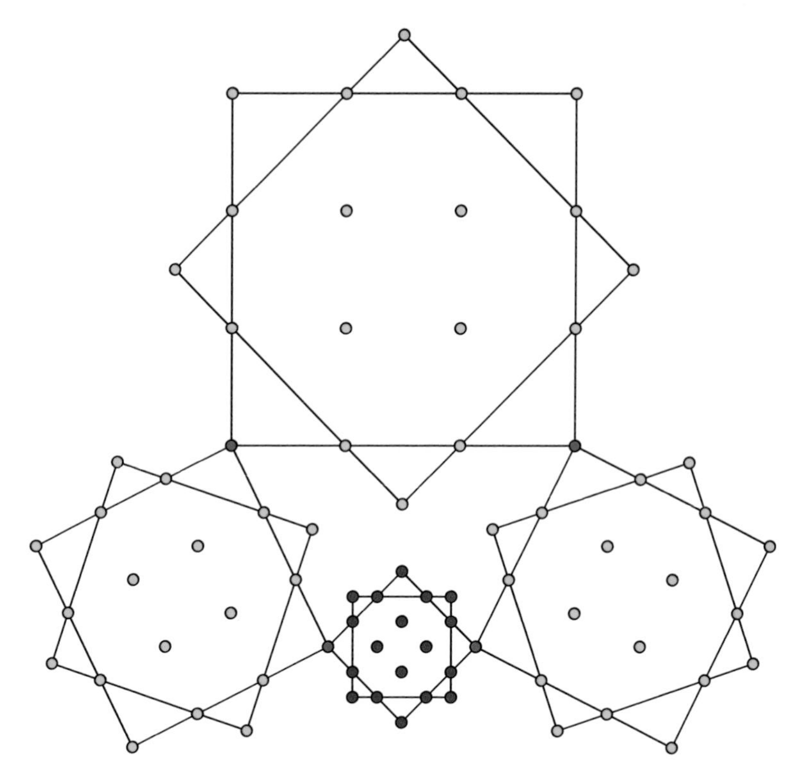

Figure 10.4. A cycle of points (blue), with each consecutive pair of blue vertices connected by a copy of the configuration from Lemma 10.15. An odd number of these copies (the yellow ones) force consecutive pairs of blue points to belong to opposite sides of any partition into two general-position subconfigurations, and the remaining copies (red) force them to belong to the same side. The configurations constructed in this way form an infinite family of obstacles to GENERAL-PARTITION$(S) < 3$ (Theorem 10.16).

problem, not-all-equal-3-satisfiability (NAE3SAT), into the problem of testing whether GENERAL-PARTITION$(S) < 3$. NAE3SAT is a problem in which the input consists of a collection of variables (each of which can be assigned to be either true or false) and a collection of clauses (each of which is a triple of terms, and each term of which is either a variable or negated variable). A valid solution to an NAE3SAT instance is a truth assignment to the variables with the property that, in every clause, at least one of the terms is true and at least one of the terms is false. The output to an NAE3SAT instance should be true if there exists a valid solution, and false otherwise. Our goal is to find a transformation from NAE3SAT instances to instances of the GENERAL-PARTITION$(S) < 3$ question that preserve this output. The use of reductions of this type is standard in the theory of NP-completeness. Whenever a problem X is known to be NP-complete, problem Y is in NP, and we can define a polynomial-time reduction from X to Y, this will prove that Y is also NP-complete. See, e.g., Garey and Johnson (1979) for details of this general proof technique, and for the fact that NAE3SAT is NP-complete.

Our reduction constructs a point for each variable, and a three-point line for each clause, with each of the three points on a clause line representing one variable or its negation. It uses a copy of Figure 10.3 between the variable point and each clause point that represents it, forcing them to be on the same side of any partition into two general position subconfigurations. Similarly, it uses more copies of Figure 10.3 between each pair of a variable point and a clause point that represents its negation, placed in a different way to force them to be on different sides of any partition. Again, if the variable and clause points are placed in sufficiently general position, there will be no extra collinearities beyond the ones in each of these subconfigurations.

Every solution to the NAE3SAT instance can be transformed into a partition of the resulting set of points into two general position subconfigurations. One of these subconfigurations consists of the variable points for variables whose truth assignment is true, clause points for these variables, clause points for negations of variables whose truth assignment is false, and the appropriate subsets of each copy of Figure 10.3. The other subconfiguration has the remaining points. Conversely, every partition into two general position subconfigurations leads to a solution to the NAE3SAT instance, obtained by setting the variables whose points belong to one of the two subconfigurations (chosen arbitrarily) to be true and setting the variables whose points belong to the other subconfiguration to be false. Therefore, this reduction does correctly preserve the outputs of its problems.

To complete the NP-completeness proof, it remains to show that this transformation from NAE3SAT instances to sets of points can be performed in polynomial time and that (to avoid representational issues) all of the points in the resulting set of points can be given small integer coordinates. First, observe that if the variable and clause points have integer coordinates that are multiples of six, then the points within each copy of Figure 10.3 will automatically also have integer coordinates. We begin the construction by placing the variable points six units apart from each other along the x-axis of the coordinate plane. Next, we place each triple of clause points, one by one, together with the subconfiguration points whose positions are determined by these clause points, in a way that avoids undesired collinearities between the newly placed points and the points that have already been placed.

We place each triple of clause points on a vertical line, with x-coordinate a multiple of six, and then (after this coordinate has been chosen) choose the y-coordinates of the three clause points on this line. If we consider any one of these points to be variable, so that its point moves continuously up and down along the clause line, then each of the associated subconfiguration points will also move along a (nonvertical) line, because its coordinates are linear combinations of the coordinates of the variable and clause points that its subconfiguration connects. We choose the x-coordinate of the clause so that no two of the lines traversed by these subconfiguration points coincide, and so that none of the lines through points of the same subconfiguration coincide with any line through two previously placed points. Each of these constraints rules out at most one possible x-coordinate for the clause line, so (if the total number of clauses and variables in the NAE3SAT instance is n) there will be $O(n^2)$ bad choices of x-coordinate. A good choice of magnitude $O(n^2)$ may be found in polynomial time by a sequential search through the possible coordinate values, testing for each value whether there is some undesired coincidence of lines.

Once the x-coordinate of the clause has been determined, we must place each of the three clause points on the vertical line with that x-coordinate, by choosing the y-coordinates for these points, again avoiding any undesired collinearities. Because of the careful choice of x-coordinate, each point in each copy of Figure 10.3 can cause a collinearity only for $O(n^2)$ choices of y-coordinates, the positions of the clause point that would cause the point to be placed on one of the lines through two previously placed points. So, again, a good choice of magnitude $O(n^2)$ may be found in polynomial time by a sequential search through the possible coordinate values, testing for each value whether there is some undesired collinearity. Repeating this process for all clauses results in a set of points in which all coordinates are integers of magnitude $O(n^2)$, constructed in polynomial time. □

Figure 10.5 shows a step of this construction, in which we place one of the clause points and its associated copy of Figure 10.3. Despite this hardness result, GENERAL-PARTITION may be fixed-parameter tractable for other parameters than its natural parameter. Corollary 10.29 will provide a result of this type.

10.4 Approximation

The greedy algorithm below approximates GENERAL-PARTITION to a square-root approximation ratio.

Algorithm 10.17 (Greedy approximation for GENERAL-PARTITION)

1. Initialize a sequence C of subconfigurations to be empty.
2. For each point p in S:
   a. If p can be added to one of the subconfigurations in C, without creating any three-point lines within that subconfiguration, then add p to the earliest such subconfiguration in the sequence.
   b. Otherwise, create a new subconfiguration consisting only of p, and add it to the end of the sequence of subconfigurations.
3. Return the approximately minimal partition C, and $|C|$ as our approximation to GENERAL-PARTITION(S).

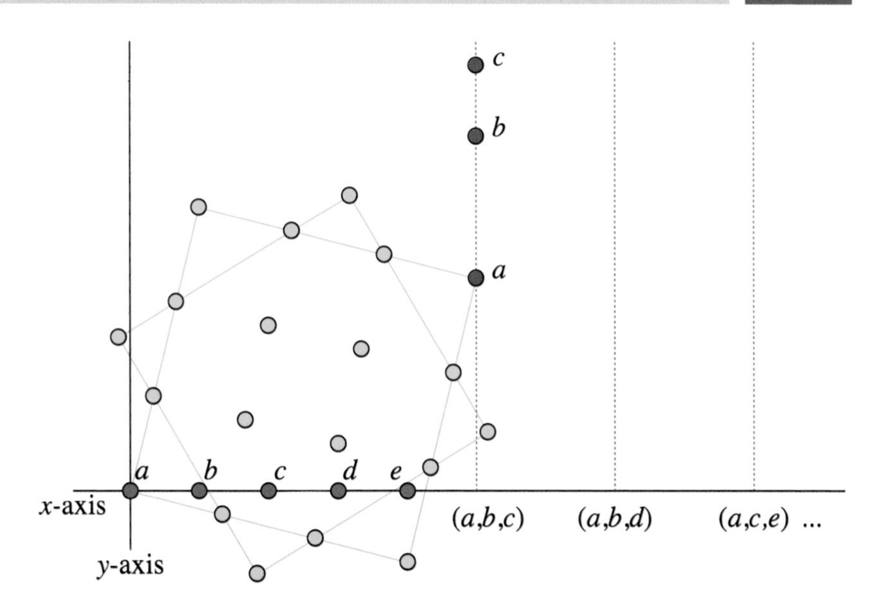

Figure 10.5. Partway through constructing an instance of GENERAL-PARTITION$(S) < 3$ from an instance of NAE3SAT, for the proof of Theorem 10.16. Five variable points (blue) have been spaced at six-unit intervals along the x-axis, the vertical line for the first clause (left dashed red line) has been fixed, and one of the three clause points on this line (red) has been placed, together with a copy of Figure 10.3 (yellow) connecting it to the corresponding variable point. The other two clause points on the same vertical line, and the lines for subsequent clauses, have yet to be determined.

Lemma 10.18

Algorithm 10.17 finds a partition into a sequence C of general-position subconfigurations with

$$|C| \leq \sqrt{2\,|S|\,\text{ONLINE}(S)}.$$

Proof. We use induction on $|S|$, with $|S| = 0$ as a base case. The first subconfiguration G in the sequence is the one produced by our greedy approximation for MAX-GENERAL, and the remaining subconfigurations are what would be produced by applying the greedy approximation for GENERAL-PARTITION to $S \setminus G$. Applying the bound of Lemma 9.15 for $|G|$, and the induction hypothesis for the remaining subconfigurations, gives

$$\begin{aligned}|C| &\leq 1 + \sqrt{2\,|S \setminus G|\,\text{ONLINE}(S \setminus G)}\\ &\leq 1 + \sqrt{2\,|S|\,\text{ONLINE}(S) - 2\,|G|\,\text{ONLINE}(S)}\end{aligned}$$

$$\leq 1 + \sqrt{2\,|S|\,\textsc{online}(S) - 2\sqrt{2\,|S|\,\textsc{online}(S)}}$$
$$\leq \sqrt{2\,|S|\,\textsc{online}(S)},$$

where the last line uses the fact that, for all $x \geq 1$, $\sqrt{x - 2\sqrt{x}} \leq \sqrt{x} - 1$. □

Corollary 10.19

For all configurations S,

$$\textsc{general-partition}(S) = O\left(\sqrt{|S|\,\textsc{online}(S)}\right).$$

Theorem 10.20

We can approximate $\textsc{general-partition}(S)$, *for a configuration S, to within an approximation ratio of* $O(\sqrt{|S|}\,)$ *in polynomial time.*

Proof. This follows immediately from the inequality

$$\textsc{general-partition}(S) \geq \frac{\textsc{online}(S)}{2}$$

and from the fact that the greedy approximation produces a partition whose number of subconfigurations is within $O(\sqrt{|S|}\,)$ of $\textsc{online}(S)$. □

Open Problem 10.21

What approximation ratio for $\textsc{general-partition}(S)$ can a polynomial-time approximation algorithm achieve?

The bound of Corollary 10.19 is tight to within a constant factor for the configurations $\textsc{line}(n)$. It is not tight for $\textsc{grid}(n, n)$: it is proportional to $n^{3/2}$, but the actual value of $\textsc{general-partition}$ is proportional to n (see Open Problem 10.1). It is even further from tight for the configurations $\textsc{ternary}(n)$, for which Observation 10.7 provides a much smaller bound on $\textsc{general-partition}$. It is consistent with all three of these examples that

$$\textsc{general-partition}(S) = O\left(\frac{\textsc{online}(S)\log|S|}{\log \textsc{online}(S)}\right),$$

but we have no proof of such a bound.

Open Problem 10.22

Is it true for all configurations S that

$$\text{GENERAL-PARTITION}(S) = O\left(\frac{\text{ONLINE}(S)\log|S|}{\log \text{ONLINE}(S)}\right)?$$

10.5 Configurations Covered by Few Lines

The main result of this section is that for large enough configurations with bounded values of LINE-COVER, GENERAL-PARTITION has a simple formula. This will allow us to compute GENERAL-PARTITION by a fixed-parameter algorithm, parameterized by LINE-COVER. To prove this result, we need a sequence of lemmas and definitions.

Lemma 10.23 (Moon and Moser, 1963)

Let A and B be sets of equal size, with a compatability relation according to which each element of A is compatible with more than half of the elements of B and each element of B is compatible with more than half of the elements of A. Then it is possible to pair the elements of A and B one-to-one so that each element of A is paired with a compatible element of B.

Definition 10.24

We define a (k, ℓ)-*ruled configuration* to be a configuration covered by k lines, each containing disjoint sets of exactly ℓ points. This partition into disjoint lines is the *ruling* of the configuration. A (k', ℓ')-*ruled subconfiguration* of a ruled configuration is a subconfiguration whose ruling uses a subset of k' of the lines, and a subset of ℓ' points on each line.

Figure 10.6 illustrates these definitions.

Lemma 10.25

Let S be a (k, ℓ)-ruled configuration, with $\ell \geq 2k^2$. Then every general-position subconfiguration G of S can be extended to a $(k, 2)$-ruled subconfiguration $H \supset G$ that remains in general position.

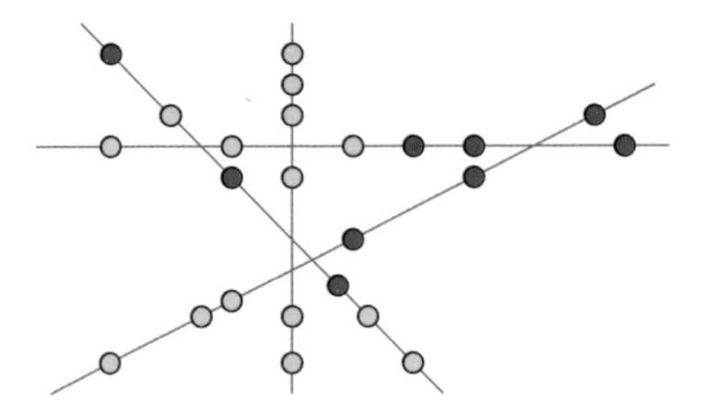

Figure 10.6. A (4, 6)-ruled configuration (red and yellow points) with a (3, 3)-ruled subconfiguration (red points only).

Proof. There are at most $2k$ points of G, determining at most $\binom{2k}{2} < 2k^2$ lines through pairs of points. So if G does not have two points on some line L of the ruling, at least one of the points on L is missed by all lines through pairs of points in G, and can be added to G preserving the general-position property of the resulting subconfiguration. Repeating this process until all lines have two points produces a $(k, 2)$-ruled subconfiguration. □

Lemma 10.26

Let n_1 and n_2 be positive integers, let S be a $(k, n_1 n_2)$-ruled configuration, and let D be a configuration disjoint from the lines of the ruling of S with $2k|D| < n_1$. Then S can be partitioned into n_1 (k, n_2)-ruled subconfigurations, such that no line through two points of the same subconfiguration passes through D.

Proof. We prove by induction on i that, when the first $i - 1$ lines of S have been partitioned into $(i - 1, n_2)$-ruled subconfigurations, the partition can be extended by one more line into a partition into (i, n_2)-ruled configurations. The result will follow by setting $i = k$.

So, suppose that we have a partition of the first $i - 1$ lines into $(i - 1, n_2)$-ruled subconfigurations. Define a set A that has n_2 elements per subconfiguration (the positions to which the points of the ith line should be assigned) and let set B be the points of the ith line of S. Both of these sets have size $n_1 n_2$. Define a compatibility relation between A, according to which a position in an $(i - 1, n_2)$-ruled subconfiguration is compatible with a point p on the ith line, if no line through p and a point of the subconfiguration passes through a point of D.

Then each point p of the ith line is compatible with the positions in all but at most $(i - 1)|D|$ subconfigurations, the ones containing points on lines through p and points of D. For each p, there are $|D|$ such lines, and each line contains points from at most $i - 1$ subconfigurations. As long as $(i - 1)|D| < n_1/2$, p will be compatible with more than half of the positions in those subconfigurations. Correspondingly, each subconfiguration determines $(i-1)n_2|D|$ lines through a point of the subconfiguration and a point of D, each of which can pass through at most one point of the ith line. As long as $(i-1)n_2|D| < n_1 n_2/2$, each position of the subconfiguration will be compatible with more than half of the points on the ith line. Both conditions will be met when $2k|D| < n_1$, as assumed.

Applying Lemma 10.23 to the sets A and B gives a compatible assignment of points of the ith line to the subconfigurations, completing the induction and proving the result. □

Lemma 10.27

For every k there exists an even number $\ell \geq 2k^2$ such that every (k, ℓ)-ruled configuration can be partitioned into $(k, 2)$-ruled subconfigurations in general position.

Proof. We use induction on k; by the induction hypothesis, let n_2 be such that every $(k-1, n_2)$-ruled configuration can be partitioned into $(k-1, 2)$-ruled subconfigurations in general position.

Choose a sufficiently large n_1, to be determined below, and let $\ell = 5n_1 n_2$. Let S be a given (k, ℓ)-ruled configuration. We let L be one of the lines of S, and partition the remaining points (arbitrarily) into five $(k-1, n_1 n_2)$-ruled subconfigurations S_i, $i \in [1, 5]$.

We will then partition each subconfiguration S_i, in turn, into $(k-1, 2)$-ruled subconfigurations in general position. We define a *dangerous point* to be a point d on L that belongs to lines through pairs of points from at least $n_1 n_2/10$ of the $(k-1, 2)$-ruled subconfigurations that we have constructed so far. Because there are at most $n_1 n_2/2$ of these subconfigurations, and each determines $4\binom{k-1}{2}$ lines (other than the lines of the ruling), there can be at most $20\binom{k-1}{2}$ dangerous points. To find these $(k-1, 2)$-ruled subconfigurations, we perform the following steps, for each i in the range $[1, 5]$.

1. Let D be the current set of dangerous points.
2. Apply Lemma 10.26 to S_i and D. This produces a partition of S_i into $(k-1, n_2)$-ruled subconfigurations, such that no line through two points of the same subconfiguration passes through D.
3. Apply the induction hypothesis to partition each of these $(k-1, n_2)$-ruled subconfigurations into $(k-2, 2)$-ruled sub-subconfigurations, each in general position.

If a point d on L becomes dangerous, it does so during the partition of a subconfiguration of S_i, with d hit by lines from fewer than $n_1 n_2/10$ $(k-1, 2)$-ruled subconfigurations earlier than S_i, at most $n_1 n_2/10$ subconfigurations within S_i itself, and none later. Thus, every point of L is hit by lines from fewer than $n_1 n_2/5$ of the $(k-1, 2)$-ruled subconfigurations, which is fewer than half of the total number $n_1 n_2/2$ of these subconfigurations.

Finally, we complete the partition by assigning the points of L to the $(k-1, 2)$-ruled subconfigurations in order to turn them into $(k, 2)$-ruled subconfigurations. To do so, make a set A with two elements per subconfiguration, representing the positions to which the points of L may be assigned. Make a set B representing the points of L. Make a compatibility relation in which a position in A is compatible with a point p in B when p can be assigned to that position's subconfiguration without creating any heavy lines. Each point in B is compatible with more than half of the elements in A, and as long as $n_1 n_2$ is sufficiently larger than $4\binom{k-1}{2}$, each position in A will be compatible with more than half of the points in B. Therefore, by Lemma 10.23, there exists a valid assignment of the points in L to general-position subconfigurations, completing the partition.

To complete the proof, it remains to calculate how large we need to set n_1 to make this proof work. There are two constraints, both of which can be satisfied for sufficiently large n_1. In order to apply Lemma 10.26, we need $n_1 > 2k|D|$, satisfied when

$$n_1 > 40k\binom{k-1}{2}.$$

And in order to apply Lemma 10.23, we need

$$\frac{n_1 n_2}{2} > 4\binom{k-1}{2},$$

which is always satisfied under the previous constraint on n_1. Therefore, the proof above is valid for all sufficiently large n_1. □

Lemma 10.27 is almost what we want but is too strict about the structure of the configurations that it covers. The main theorem of this section generalizes this result to arbitrary configurations.

Theorem 10.28

For every k there exists n such that every configuration S with LINE-COVER$(S) \le k$ *and* $|S| \ge n$ *has* GENERAL-PARTITION$(S) = \lceil$ONLINE$(S)/2\rceil$.

Proof. The configurations S with LINE-COVER$(S) \le k$ and $|S| \ge n$ also have ONLINE$(S) \ge n/k$, so it is equivalent to state the theorem as: for every k there exists ℓ such that every configuration S with LINE-COVER$(S) \le k$ and ONLINE$(S) \ge \ell$ has GENERAL-PARTITION$(S) = \lceil$ONLINE$(S)/2\rceil$. Let ℓ' be the corresponding number from Lemma 10.27; we will show that it suffices to set $\ell = \ell' + 2\binom{k}{2}$.

Let S be an arbitrary configuration with LINE-COVER$(S) \leq k$ and ONLINE$(S) \geq \ell$. Let L_i be a set of k lines that cover S. By adding dummy points if necessary to S, we can assume without loss of generality that all lines L_i contain exactly ONLINE(S) points. We may apply a greedy process similar to that in Lemma 10.25 to find, for each point p on more than one line L_i, a general-position subconfiguration that includes p and has two points on each line L_i. Removing these subconfigurations from S reduces to the case where the lines L_i are disjoint and where each of them has at least ℓ' lines. We may then apply Lemma 10.25 repeatedly to find and remove general-position subconfigurations from S, reducing to the case where each L_i has exactly ℓ' lines. The result follows from Lemma 10.27. □

Corollary 10.29

Computing GENERAL-PARTITION(S) *is fixed-parameter tractable when parameterized by* LINE-COVER(S).

Proof. Given a configuration S, let $k =$ LINE-COVER(S) and let n be the number given for k by Theorem 10.28. When $|S| \geq n$, we can compute GENERAL-PARTITION directly by the theorem, and in the remaining case we can compute it by a brute force search in an amount of time that can be bounded by a function of k. □

Multiplication by a power of k at each level of the induction causes the bound on n in Theorem 10.28, and the size of the kernel in the kernelization of Corollary 10.29, to be exponential in a polynomial of k.

Open Problem 10.30

What is the tightest possible dependence of n on k in Theorem 10.28? Does GENERAL-PARTITION, parameterized by LINE-COVER, have a polynomially sized kernel?

11 Convexity

A configuration is in *convex position* if it forms the vertices of a convex polygon. By Carathéodory's theorem,[1] a configuration that is not in convex position has one point contained in the convex hull of two or three other points. Thus, there are two obstacles to convexity: LINE(3) and a triangle with an interior point (Figure 3.3, which we denote as TETRAD).

Definition 11.1

We define the properties and parameters

$$\begin{aligned} \text{CONVEX} &= \text{FORBIDDEN}\big(\text{LINE}(3), \text{TETRAD}\big) \\ \text{MAX-CONVEX} &= \text{AVOIDS}\big(\text{LINE}(3), \text{TETRAD}\big) \\ \text{DELETE-TO-CONVEX} &= \text{HITTING}\big(\text{LINE}(3), \text{TETRAD}\big) \\ \text{CONVEX-PARTITION} &= \text{PARTITION}\big(\text{LINE}(3), \text{TETRAD}\big). \end{aligned}$$

Arkin et al. (2003) called CONVEX-PARTITION(S) the *convex cover number*.

11.1 Happy Endings Revisited

Parameter MAX-CONVEX is the central object of study in the happy ending problem. We now fulfill our promise in Chapter 1 to describe a configuration of 2^{k-1} points with MAX-CONVEX $= k$, for every $k \geq 0$.[2]

[1] Carathéodory (1907).

[2] In this description, we follow Morris and Soltan (2000), who in turn follow Erdős and Szekeres (1960), Lovász (1979), and Kalbfleisch and Stanton (1995).

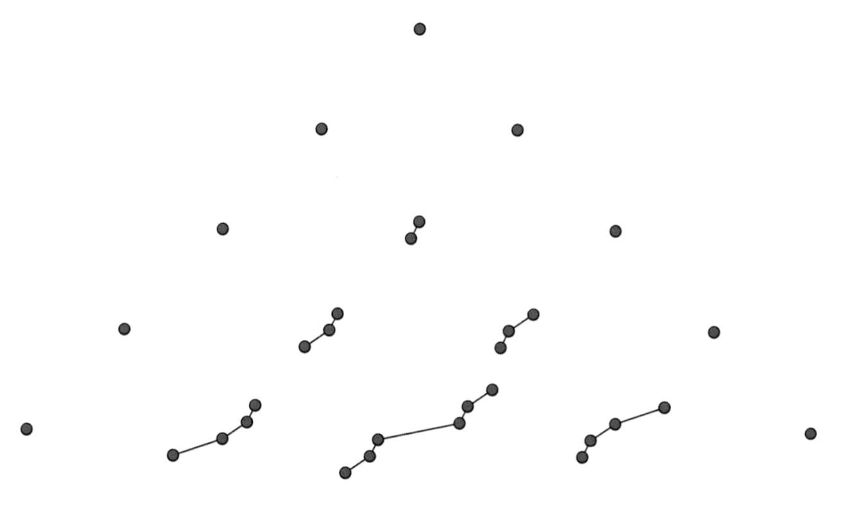

Figure 11.1. A Pascal triangle of configurations.

To do so, we construct an analogue of Pascal's triangle for point sets. Pascal's triangle is the triangular array

$$\begin{array}{ccccccccccccc}
 & & & & & & 1 & & & & & & \\
 & & & & & 1 & & 1 & & & & & \\
 & & & & 1 & & 2 & & 1 & & & & \\
 & & & 1 & & 3 & & 3 & & 1 & & & \\
 & & 1 & & 4 & & 6 & & 4 & & 1 & & \\
 & 1 & & 5 & & 10 & & 10 & & 5 & & 1 & \\
1 & & 6 & & 15 & & 20 & & 15 & & 6 & & 1 \\
\vdots & & \vdots & & \vdots & & \vdots & & \vdots & & \vdots & & \vdots
\end{array}$$

in which each number is the sum of the two numbers above it to its left and right. The number in position b of row a (starting from zero) is a *binomial coefficient* denoted $\binom{a}{b}$. For instance, $\binom{4}{2} = 6$, the middle number of row four.

Analogously, build a triangular array of configurations (Figure 11.1), each realized by a path monotone in both the x- and y-directions. Each path glues together the two smaller paths above it to its left and right, spaced widely enough apart so that all lines through two left points pass above the right path, and all lines through two right points pass below the left path. By analogy to the binomial coefficients, let the configuration in position b of row a be called BINOMIAL(a, b).

Then BINOMIAL(a, b) has size $\binom{a}{b}$, but we promised a configuration whose size is a power of two. To achieve this, glue together all the configurations on row a

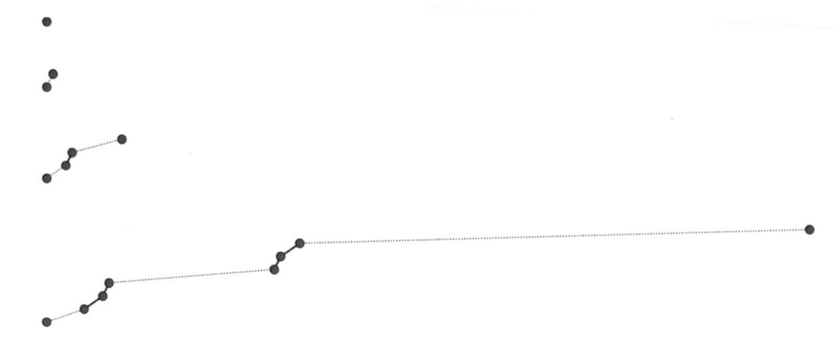

Figure 11.2. Gluing together the configurations BINOMIAL(a, b) from a single row a to form the configurations PASCAL(a).

of the triangle, in left-to-right order, again making the horizontal spacing in each gluing step wide enough so lines through pairs of points on the left pass above the right side and the lines through pairs on the right pass below the left. Figure 11.2 shows four examples of the resulting configuration, which we call PASCAL(a), for $a \in \{0, 1, 2, 3\}$. PASCAL(a) has size

$$\sum_{b=0}^{a} \binom{a}{b} = 2^a.$$

To analyze the convex subconfigurations of these configurations, we consider two special types of convex sets of points. A *cup* is a finite subset of the graph of a convex function; all of its points lie on or below the line through its leftmost and rightmost point. A *cap* is an upside-down cup: a finite subset of the graph of a concave function, with all points on or above the line through its leftmost and rightmost point. The point sets BINOMIAL(a, 1) on the left side of the triangle (next to the single points BINOMIAL(a, 0)) are cups, and the point sets BINOMIAL(a, $a-1$) on the right side are caps.

Observation 11.2

When two monotone paths A and B are glued together using the gluing operation that we used in constructing the BINOMIAL *and* PASCAL, *the only convex subconfigurations of the result that include points from both A and B have the form of a cap in A and a cup in B.*

Lemma 11.3

In the monotone path realizing BINOMIAL(a, b), *the largest cap has size* $b+1$ *and the largest cup has size* $a-b+1$.

Proof. For $b = 0$ or $a = b$, the result follows immediately from the fact that in those cases $|\text{BINOMIAL}(a, b)| = 1$. For other b, if the largest cup or cap were entirely contained in one of the two subconfigurations $\text{BINOMIAL}(a-1, b-1)$ and $\text{BINOMIAL}(a-1, b)$ out of which $\text{BINOMIAL}(a, b)$ is formed, the result would follow by induction on that subconfiguration. By Observation 11.2, any other cap must consist of a cap in $\text{BINOMIAL}(a-1, b-1)$ and a single point in $\text{BINOMIAL}(a-1, b)$. Symmetrically, any other cup must consist of a single point in $\text{BINOMIAL}(a-1, b-1)$ and a cup in $\text{BINOMIAL}(a-1, b)$. The result follows by applying the induction hypothesis and adding one. □

Lemma 11.4

For $a > 0$, $\text{MAX-CONVEX}\big(\text{BINOMIAL}(a, b)\big) \le a$.

Proof. As before this follows immediately for $b = 0$ or $a = b$. For other b, if the largest convex subconfiguration were entirely contained in one of the two subconfigurations out of which $\text{BINOMIAL}(a, b)$ is formed, the result would follow by induction. Otherwise, by Observation 11.2, any other convex subconfiguration must consist of a cap in the left subconfiguration and a cup in the right subconfiguration. The result follows by using Lemma 11.3 to bound the sizes of this cup and cap. □

Theorem 11.5

$\text{MAX-CONVEX}\big(\text{PASCAL}(a)\big) = a + 1$.

Proof. We can find a convex set of this size in the parts of PASCAL coming from $\text{BINOMIAL}(a, 0)$ and $\text{BINOMIAL}(a, 1)$, so it remains to prove that no larger convex set exists. By Observation 11.2, a convex subconfiguration of $\text{PASCAL}(a)$ must be either a subconfiguration of $\text{BINOMIAL}(a, b)$ for some b or a cap in $\text{BINOMIAL}(a, b)$, at most one point in each of $\text{BINOMIAL}(a, b+1)$, $\text{BINOMIAL}(a, b+2), \ldots, \text{BINOMIAL}(a, c-1)$, and a cup in $\text{BINOMIAL}(a, c)$, for some b and c with $0 \le b < c \le a$. In the first case, the result follows from Lemma 11.4, and in the second case it follows from Lemma 11.3. □

There has been some research on MAX-CONVEX in general-position configurations with the property $\text{FORBIDDEN}(X)$, for different choices of X. For some choices of X, such as the regular pentagon plus its center point, there still exist configurations with X forbidden but with MAX-CONVEX at most logarithmic. That is, forbidding X does not significantly change how small MAX-CONVEX can be. But for some other choices of X, forbidding X can cause MAX-CONVEX to grow more quickly, as a polynomial function of the size of the configuration.[3] We will see an example of this phenomenon in Corollary 14.14.

11.2 Additional Examples

We can completely classify the configurations S for which MAX-CONVEX(S) < 4, as a refinement of the known classification of configurations that have no empty quadrilateral.[4]

Observation 11.6

The configurations with MAX-CONVEX(S) < 4 *are the following.*

- LINE(n), *for which* MAX-CONVEX(S) < 3.
- *The configurations formed as* LINE(n) *plus one point off the line.*
- *The configurations formed from* LINE(n) *by adding one point on each side of the line, such that the segment between the two added points does not separate any pair of points from the* LINE(n) *subconfiguration.*
- *The two six-point configurations labeled* 111 *and* 222 *in Figure 8.6.*

Thus, arbitrarily large configurations can have MAX-CONVEX(S) < 4, but these configurations have a very special form with many points on a line.

Next, we consider the grids. A classical theorem of Jarník (1926) allows us to bound the largest size of a convex subconfiguration of any grid. This theorem states that a convex curve of length L can contain at most

$$\frac{3}{\sqrt[3]{2\pi}}L^{2/3} + O(L^{1/3})$$

points with integer coordinates. Since a convex curve in GRID(n, n) has length at most $4(n-1)$, it follows that MAX-CONVEX$\big($GRID(n, n)$\big) = O(n^{2/3})$. The proof of this bound generalizes to any grid.

Observation 11.7

For any positive n and m,

$$\text{MAX-CONVEX}\big(\text{GRID}(m, n)\big) = O\big(\min(m, n, (mn)^{1/3})\big).$$

[3] Károlyi and Solymosi (2006); Károlyi and Tóth (2012). [4] Eppstein (2010).

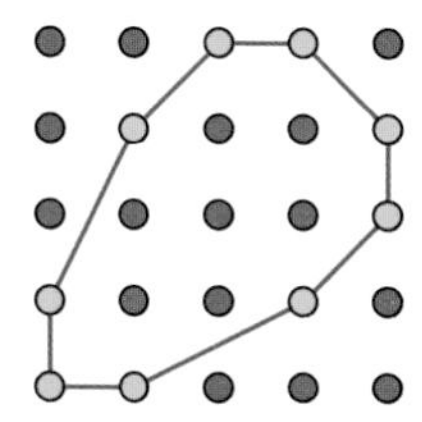

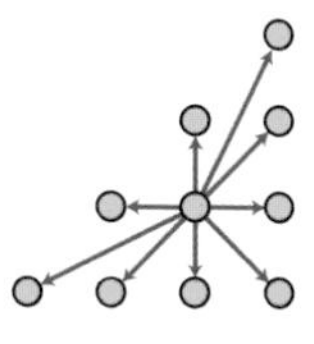

Figure 11.3. Left: MAX-CONVEX(GRID$(5,5)) = 9$. Right: the difference between each point and its clockwise neighbor in this nine-point convex set.

Proof. The parts of this bound stating that a convex subset of the grid can have at most $O(\min(m, n))$ points follow from the fact that each axis-parallel line through the grid can contain only two points of any convex subset.

The $O((mn)^{1/3})$ bound comes from the fact that, in a convex subset of GRID(m, n) with k points, most of the convex hull edges must connect pairs of points whose x-coordinates differ by $O(m/k)$, for otherwise the convex hull would be too wide. Similarly, most of the edges must connect pairs of points whose y-coordinates differ by $O(n/k)$. Therefore, the slopes of most of the edges are fractions a/b where the numerator a and denominator b are both small: $a = O(n/k)$ and $b = O(m/k)$. But there can be only $O(mn/k^2)$ distinct slopes made from numbers this small, and at most two convex hull edges can have each slope. Therefore, $k = O(mn/k^2)$. Eliminating k from the right side of this bound gives $k = O((mn)^{1/3})$, as claimed. □

Figure 11.3 shows MAX-CONVEX(GRID$(5,5))$, and the vectors of differences between consecutive points on the convex hull of this convex set. As in the proof of Observation 11.7, these vectors lie in a small grid.

Because each convex polygon in a grid must be small, the number of polygons in a convex partition must be large: CONVEX-PARTITION(GRID$(n, n)) = \Omega(n^{4/3})$. A more precise result, showing the existence of a partition into this many nested convex subsets, is given in Theorem 12.7.

As well as the grids, it is also of interest to consider the MAX-CONVEX parameter for the sawtooth configurations.

Observation 11.8

For any n,

$$\text{MAX-CONVEX}(\text{SAWTOOTH}(n)) = \left\lfloor \frac{3n}{2} \right\rfloor.$$

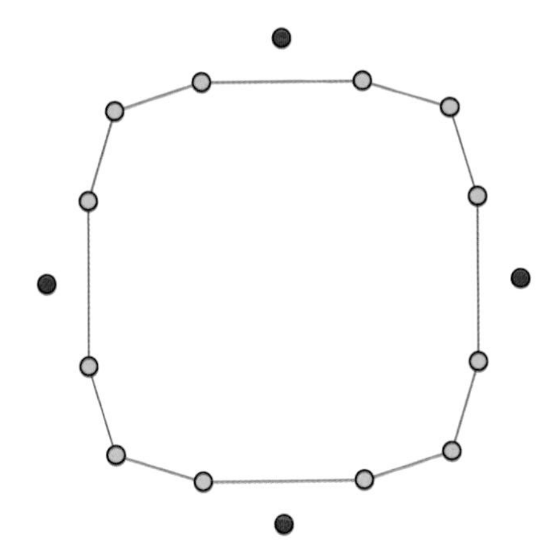

Figure 11.4. Removing $\lceil n/2 \rceil$ convex hull vertices (red) from a sawtooth configuration (here, SAWTOOTH(8)) leaves a maximum convex subconfiguration with $\lfloor 3n/2 \rfloor$ points (here, 12 points).

Proof. Recall that SAWTOOTH(n) is formed by a convex polygon together with equally many points near the polygon edge midpoints. A convex subconfiguration with more than three points cannot include both endpoints and the near-midpoint of a single edge, because the near-midpoint is interior to the convex hull of these three points with any fourth point of the configuration. Therefore, a large convex subconfiguration must omit at least one out of every four consecutive points in cyclic order around the sawtooth. If n is even, the $2n$ points of the sawtooth can be grouped into $n/2$ consecutive quadruples, so at least $n/2$ points must be omitted. If n is odd, then after omitting any one point the remaining $2n-1$ points can be grouped into $(n-1)/2$ consecutive quadruples (with one point left over), so $(n+1)/2$ points must be omitted. Therefore, MAX-CONVEX can be at most the stated value.

To show that MAX-CONVEX is also at least the stated value, remove from SAWTOOTH(n) a subset of $\lceil n/2 \rceil$ vertices of its convex hull, so that every convex hull edge has one of its endpoints removed. The remaining subconfiguration is convex, and has the stated size. □

Figure 11.4 illustrates the case $n = 8$ of Observation 11.8.

Corollary 11.9

Sample-based property testing for convexity requires samples of size $\Omega(n^{2/3})$ to achieve a constant false positive rate p for configurations that are $1/4$-far from being convex.

Proof. SAWTOOTH(n) is 1/4-far from convex, and a sample-based property testing algorithm, when run on SAWTOOTH(n), will generate a false positive unless its sample includes the two endpoints and near-midpoint of at least one edge of the convex hull. The argument that a sample size of $\Omega(n^{2/3})$ is needed to have a good probability of including at least one such triple is the same as for Theorem 9.9, so we omit the details. □

On the other hand, Theorem 6.8 proves that sample size $O(n^{3/4})$ works, but does not match the lower bound of Corollary 11.9.

Open Problem 11.10

Does the sample-based property testing algorithm for convexity, with sample size $O(n^{2/3})$, achieve constant false positive rate, or is sample size $\Omega(n^{3/4})$ needed?

11.3 Obstacles

By Observation 7.5 and Theorem 7.9, the two parameters MAX-CONVEX(S) and DELETE-TO-CONVEX(S) both have polynomial obstacle size. However, this is not true for CONVEX-PARTITION(S).

Theorem 11.11

The parameter CONVEX-PARTITION(S) *has an infinite number of obstacles for* CONVEX-PARTITION(S) < 3.

Proof. Consider the configuration S_n shown in Figure 11.5. It has a very similar structure to SAWTOOTH(n), with the following two differences:

- We require that the number of vertices on the convex hull (the parameter n in the notation S_n) be an odd number.
- Instead of placing one point near the midpoint of each convex hull edge, we place two, one inside the other (along a radius through the centroid of the configuration). These two points should be placed close enough to each other, and close enough to the midpoint of the convex hull edge, so that any line through two points near two consecutive convex hull edges will separate one convex hull vertex from the others.

S_n has the property that, for each pair (p, q) of points near the same convex hull edge, no convex quadrilateral has p and q as diagonally opposite vertices.

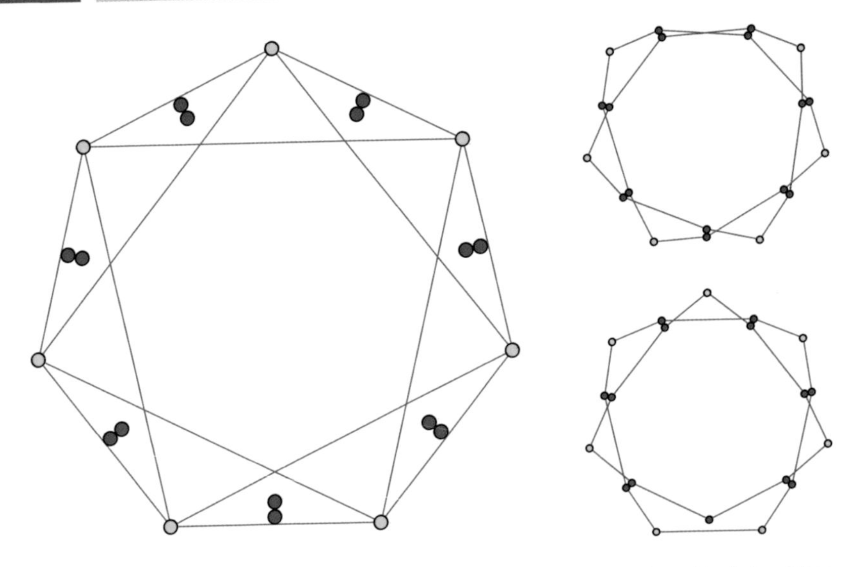

Figure 11.5. An obstacle for CONVEX-PARTITION(S) < 3, as described by Theorem 11.11. The two smaller images on the right show that, if any point is removed, the remaining points can be partitioned into two convex polygons.

For, any other two potential quadrilateral vertices can be separated from p and q by a line. Therefore, any convex subconfiguration containing both p and q can be separated by line pq from half of the points in S_n, showing that the points outside the configuration are not convex. Thus, if S_n is to be divided into two convex subconfigurations, each of these two subconfigurations must contain exactly one point from each pair (p, q).

However, if one of the subconfigurations in such a partition contains a point p from one of the pairs (p, q), it cannot contain both of the endpoints of the nearby convex hull edge. For, if it did contain both of them, it would not be able to include any other points from S_n while remaining convex, and therefore the other of the two subconfigurations would not be convex. Thus, if S_n is to be divided into two convex subconfigurations, neither subconfiguration can contain two consecutive convex hull vertices from S_n.

Because n is odd, the cyclic sequence of convex hull vertices cannot be partitioned into two subsets, neither of which contains two consecutive convex hull vertices. Thus, S_n has no partition into two convex subconfigurations. That is, CONVEX-PARTITION(S_n) ≥ 3.

However, if any convex hull vertex is removed from S_n, the remaining subconfiguration S'_n has CONVEX-PARTITION(S'_n) $= 2$. For, in this case, we may form a partition into two convex polygons by assigning the remaining convex hull vertices alternatingly to the two polygons, and choosing arbitrarily for each pair of interior points of S_n which of the two points should belong to which of the two polygons (Figure 11.5, upper right).

Also, if any interior point is removed from S_n, the remaining subconfiguration S''_n again has CONVEX-PARTITION$(S''_n) = 2$. For, in this case, we may assign the two nearby convex hull vertices to a single polygon of the partition, and then alternate between the two polygons for the remaining convex hull vertices. The interior point near the removed one should go into the polygon that does not include the two nearby convex hull vertices, and otherwise each pair of interior points may be partitioned arbitrarily between the two polygons. Again, the result is a partition into two convex polygons (Figure 11.5, lower right).

Thus, each configuration S_n has CONVEX-PARTITION$(S_n) \geq 3$ and is minimal for these parameter values, so it forms an obstacle. There are infinitely many odd numbers to choose for the value of n, so there are infinitely many obstacles. □

11.4 Finding Convex Subconfigurations

We may test whether a configuration is CONVEX by constructing its convex hull in time $O(n \log n)$ (see Section 6.1) and then checking whether the hull vertices include all of the given points.

As Chvátal and Klincsek (1980) first showed, MAX-CONVEX is also computable in polynomial time. They used *dynamic programming*, a technique in which one solves an optimization problem by formulating and solving overlapping subproblems, each of which can be solved by combining solutions to simpler subproblems. To simplify the description of this algorithm, it is helpful to start by finding a restricted type of convex subconfiguration, one in which a particular convex hull vertex is always included.

Algorithm 11.12 (Largest convex subconfiguration that includes a convex hull vertex p)

We assume that we are given as input a configuration S, and a point p of the convex hull of S. We will construct, for each other point q and each line ℓ through q and another point, a convex polygon $C(q, \ell)$. This polygon is required to have pq as one of its edges, with p clockwise of q, and to be entirely on one side of line ℓ. Among polygons that obey these constraints, it should have the largest possible number of points. To compute these polygons, we perform the following steps.

1. Sort the points of S into clockwise order around p, starting from the clockwise neighbor of p along the convex hull and ending with the counterclockwise neighbor.
2. For each point q in the resulting sorted order, do the following.
   a. Sort the lines qr that pass through q and another point r, into clockwise order around q, starting from line pq.
   b. For each line ℓ in this sorted order, perform the following steps.

i. If $\ell = pq$, set $C(q, \ell) = \{p, q\}$, and go on to the next line in the sorted sequence of lines.
ii. Let ℓ' be the line immediately before ℓ in the sorted order around q, and set $C(q, \ell) = C(q, \ell')$.
iii. For each point r on ℓ that precedes q in the sorted order around p, let ℓ' be the line immediately before ℓ in the sorted order around r. Consider the convex polygon $P = C(r, \ell') \cup \{q\}$. If P has more points than the current assignment to $C(q, \ell)$, change $C(q, \ell)$ to equal P.

3. Return the largest polygon $C(q, \ell)$ found in the above search.

We can then apply Algorithm 11.12 repeatedly to find the overall largest convex subconfiguration of a given configuration S.

Algorithm 11.13 (Finding the largest convex subconfiguration)

1. While the input configuration S is nonempty, do the following.
   a. Compute the convex hull of S, and let p be an arbitrary vertex of the convex hull.
   b. Apply Algorithm 11.12 to find the largest convex subconfiguration of S that includes p.
   c. Remove p from S.
2. Return the largest convex subconfiguration found in the above search.

As described above, and with some care with implementation details, Algorithm 11.13 can be implemented to run in time $O(|S|^3 \log |S|)$ and space $O(|S|^2)$. Edelsbrunner and Guibas (1989) reformulated Algorithm 11.12 as a search through a dual arrangement of lines, allowing them to apply their topological sweeping method to this subproblem. Using it, they proved the following.

Theorem 11.14 (Edelsbrunner and Guibas, 1989)

Given a configuration S, we can compute MAX-CONVEX(S) *in time $O(|S|^3)$ and space $O(|S|)$.*

Given this calculation of MAX-CONVEX, we can then calculate

$$\text{DELETE-TO-CONVEX}(S) = |S| - \text{MAX-CONVEX}(S),$$

allowing DELETE-TO-CONVEX to be computed in the same time bounds.

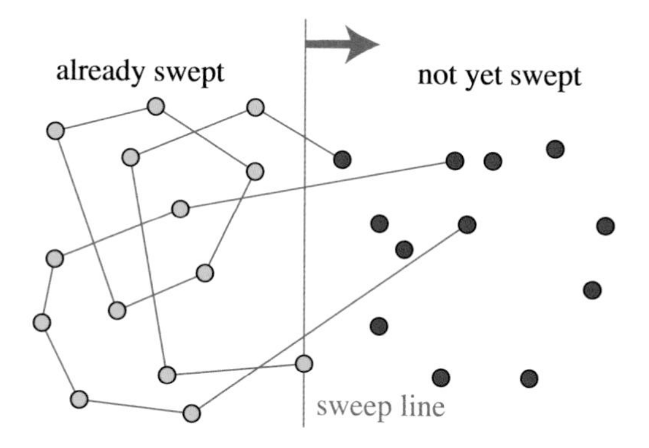

Figure 11.6. Partway through testing CONVEX-PARTITION$(S) < k$ (Theorem 11.15). Three partial convex polygons that together cover all of the points to the left of the sweep line are shown. The way these polygons cross the sweep line constitutes a state of the algorithm, and the algorithm considers all possible states simultaneously (not just the one shown).

11.5 Finding Convex Partitions

Finding a partition into a minimum number of vertex-disjoint convex polygons is NP-hard, by a reduction from 1-in-3 SAT.[5] However, this reduction only proves that instances with large numbers of polygons are hard. When the number of polygons is bounded, the problem is polynomial.

Theorem 11.15

For any fixed k, testing whether CONVEX-PARTITION$(S) < k$ *can be done in polynomial time.*

Proof. We follow a plane-sweep strategy, in which we sweep a vertical line from left to right over a set of points representing the given configuration S and keep track of the possible partitions of the points on or to the left of the sweep line (Figure 11.6). (For points given by their order type rather than by coordinates, we instead sweep radially around a vertex of the convex hull; this alternative sweep order does not affect the rest of the algorithm.)

We define a *state* of this sweep to be a description of how a set of k convex polygons intersects the sweep line. A convex polygon can intersect this line along a vertical edge, at a single vertex, or at two points, either of which might be a point from S or interior to a polygon edge. Alternatively, one of the convex polygons might be disjoint from the sweep line, which we treat as two distinct

[5] Arkin et al. (2003).

possible states: it can be entirely to the left or right of the sweep line. Therefore, a single convex polygon can intersect the sweep line in $O(n^2)$ different states (where $n = |S|$). The collection of (fewer than k) convex polygons can intersect it in $O(n^{2k-2})$ different states.

Two positions of the sweep line are considered equivalent if they form the same left-right partition of S, and distinct otherwise. For each of the $O(n)$ distinct positions p of the sweep line, and each of the $O(n^{2k})$ states s that the partition into fewer than k convex polygons can form with respect to position p, we determine whether a collection of convex polygons in state s can cover all of the points on or to the left of the sweep line. This test can be performed by considering the most immediate previous position of the sweep line, and the states from that position that could have led to any of the states in position p. The whole problem has CONVEX-PARTITION$(S) < k$ if and only if this algorithm reaches a position where the sweep line has passed the rightmost point of S, and the (unique) state for this position can still cover all of the points to the left of the sweep line.

The time for this sweep is $n^{O(k)}$, polynomial when k is fixed. □

Because the time bound of the algorithm described above has an exponent that varies with k, the algorithm is not fixed-parameter tractable.

Open Problem 11.16

Is there a fixed-parameter tractable algorithm for computing CONVEX-PARTITION?

The hardness reduction for partitioning into convex polygons expands the 1-in-3 SAT instance that it starts from by a polynomial factor. Because of this expansion, it does not show hardness of approximation for CONVEX-PARTITION.

Observation 11.17 (Arkin et al., 2003)

CONVEX-PARTITION *can be approximated to within a logarithmic approximation ratio.*

Proof. We apply a greedy algorithm that repeatedly finds and removes the convex polygon with the largest possible number of vertices, using the algorithm from Section 11.4. If the configuration S has CONVEX-PARTITION$(S) = k$, then each removal will reduce the size of the remaining subconfiguration by a factor of $(1 - 1/k)$ or smaller. At most $O(k \log n)$ reductions can be performed to a configuration of initial size n before reducing it to the empty configuration, giving the approximation ratio. □

Open Problem 11.18

Is CONVEX-PARTITION APX-hard? Does it have a better than logarithmic approximation ratio?

11.6 Inequalities

Both the original happy ending theorem of Erdős and Szekeres and Suk's strengthening of it imply that, for the purposes of parameterized complexity, MAX-CONVEX is equivalent to MAX-GENERAL and LINE-COVER.

Observation 11.19

$\textsc{max-convex} \doteq \textsc{max-general} \doteq \textsc{line-cover}$.

Proof. $\textsc{max-convex}(S) \leq \textsc{max-general}(S)$ for all S, because every convex subconfiguration is in general position. By the happy ending theorem,

$$\textsc{max-convex}(S) = \Omega\big(\log(\textsc{max-general}(S))\big).$$

Therefore, $\textsc{max-convex} \doteq \textsc{max-general}$. The equivalence to LINE-COVER follows from Observation 9.10. □

More generally, we can use the happy ending theorem to determine which parameters of the form $\textsc{avoids}(C_1, C_2, \ldots)$ are equivalent to SIZE, and which are smaller.

Theorem 11.20

Let $C_1, C_2, \ldots$ be any set of obstacles. Then

$$\textsc{avoids}(C_1, C_2, \ldots) \ll \textsc{size}$$

if and only if at least one of the obstacles C_i is a LINE *or* POLYGON *configuration. Otherwise, when neither of these two types of obstacles is included as one of the C_i,*

$$\textsc{avoids}(C_1, C_2, \ldots) \doteq \textsc{size}.$$

Proof. If $C_i = \text{LINE}(j)$, then the configurations $\text{LINE}(n)$ have unbounded size ($|\text{LINE}(n)| = n$) but bounded values of the AVOIDS parameter:

$$\text{AVOIDS}\big(C_1, C_2, \ldots; \text{LINE}(n)\big) \leq j - 1.$$

Similarly, if $C_i = \text{POLYGON}(j)$, then the configurations $\text{POLYGON}(n)$ have unbounded size ($|\text{POLYGON}(n)| = n$) but bounded values of the AVOIDS parameter:

$$\text{AVOIDS}\big(C_1, C_2, \ldots; \text{POLYGON}(n)\big) \leq j - 1.$$

Thus, in either case, we have

$$\text{AVOIDS}(C_1, C_2, \ldots) \ll \text{SIZE}$$

as claimed.

On the other hand, suppose that neither lines nor polygons are included among the obstacles C_i. In this case, for all configurations S,

$$\text{AVOIDS}(C_1, C_2, \ldots; S) \geq \max\big(\text{ONLINE}(S), \text{MAX-CONVEX}(S)\big)$$

because both the largest line in S and the largest polygon in S avoid the given obstacles. However, by the happy ending theorem, $\text{MAX-CONVEX} \doteq \text{MAX-GENERAL}$, and by Observation 9.12,

$$\max(\text{ONLINE}, \text{MAX-GENERAL}) \doteq \text{SIZE}.$$

Combining these relations gives

$$\text{AVOIDS}(C_1, C_2, \ldots) \doteq \text{SIZE}. \qquad \square$$

For an application of Theorem 11.20, see Observation 12.2.

DELETE-TO-CONVEX and CONVEX-PARTITION obey the following relations with each other and with the previously defined parameters.

Observation 11.21

$$\text{DELETE-TO-GENERAL} \ll \text{DELETE-TO-CONVEX}$$

and

$$\text{GENERAL-PARTITION} \ll \min(\text{DELETE-TO-GENERAL}, \text{CONVEX-PARTITION}).$$

Proof. For all configurations S, the relations

$$\text{DELETE-TO-GENERAL}(S) \leq \text{DELETE-TO-CONVEX}(S)$$

and

$$\text{GENERAL-PARTITION}(S) \leq \text{CONVEX-PARTITION}(S)$$

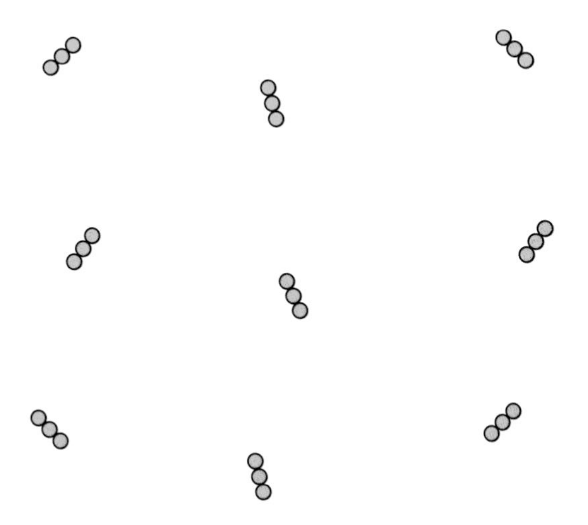

Figure 11.7. Replacing grid points by triples of collinear points that are otherwise in general position produces a configuration S with GENERAL-PARTITION(S) $= 2$ but DELETE-TO-GENERAL(S) and CONVEX-PARTITION(S) unbounded (Observation 11.21).

follow immediately from the fact that a configuration in convex position is necessarily in general position. The inequality

$$\text{GENERAL-PARTITION}(S) \leq 1 + \left\lceil \frac{\text{DELETE-TO-GENERAL}(S)}{2} \right\rceil$$

is an instance of the partition-removal inequality in Observation 5.15. For a family of configurations for which DELETE-TO-GENERAL is bounded but for which DELETE-TO-CONVEX is unbounded, perturb GRID(n, n) to be in general position without changing the orientation of any triples of points that were not initially collinear.

For configurations for which GENERAL-PARTITION is bounded but for which both of the parameters DELETE-TO-GENERAL and CONVEX-PARTITION are unbounded, replace the points of GRID(n) by triples of collinear points, otherwise in general position with respect to each other (Figure 11.7). A partition of this configuration into two general-position subconfigurations may be obtained by assigning one point of each triple to one subconfiguration and the other two points of each triple to the other subconfiguration, arbitrarily. However, finding a single general-position subconfiguration requires the removal of at least one point from each triple, so DELETE-TO-GENERAL(S) is proportional to the total number of points. And in any convex subconfiguration, only the two most extreme grid columns of the subconfiguration can include points from more than two triples, so every such subconfiguration has a number of points at most proportional to the number of grid rows and columns; therefore, CONVEX-PARTITION(S) is also at least proportional to the number of grid rows and columns. □

11.7 Well-Quasi-Ordering

Theorem 11.22

The configurations S for which MAX-CONVEX(S) < 5 *are well-quasi-ordered, and the number of such configurations with n points is polynomial in n. Any monotone parameter of these configurations is nonuniform fixed-parameter tractable. However, the configurations for which* MAX-CONVEX(S) < 7 *are not well-quasi-ordered, and the number of such configurations is exponential in* $n \log n$.

Proof. For the results on configurations with MAX-CONVEX(S) < 5, see Section 17.6; for the results on configurations with MAX-CONVEX(S) < 7, see Section 15.4. □

Open Problem 11.23

Are the configurations with MAX-CONVEX(S) < 6 well-quasi-ordered, and how many such configurations are there?

Theorem 11.24

The configurations S for which DELETE-TO-CONVEX(S) < 2 *are well-quasi-ordered. However, this is not true of the configurations for which* DELETE-TO-CONVEX(S) < 3.

Proof. The configurations with DELETE-TO-CONVEX(S) $= 0$ are just the ones of the form POLYGON(n). When DELETE-TO-CONVEX(S) $= 1$ and the configuration can be made convex by removing a point p outside the convex hull of the remaining points, its order type can be determined by sweeping a line radially through p and recording, at each point of the sweep where this line crosses one of the other points of S, whether the crossed point is on the arc of the convex hull nearest to p, the arc farthest from p, or whether the line simultaneously crosses points on both arcs. The sequences of positions recorded in this way are well-ordered by Higman's lemma. Similarly, when the removed point p lies within the convex hull, its order type can be determined by sweeping a line through p in slope order and recording whether each point that the line sweeps through is on the left side of p, the right side of p, or whether the line sweeps

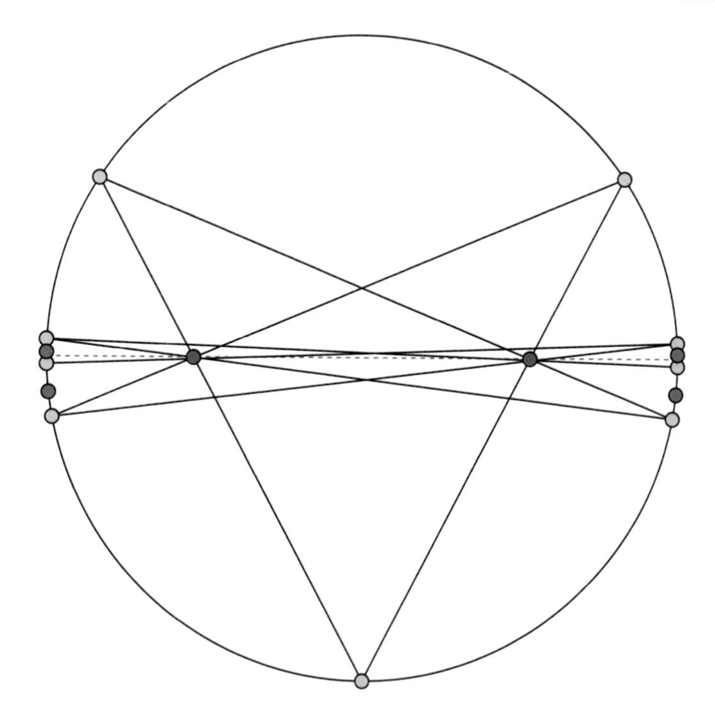

Figure 11.8. One of the configurations in an antichain of configurations with DELETE-TO-CONVEX(S) < 3.

through a point on each side simultaneously. The sequences recorded in this way are well-ordered, and the ordering on configurations is a refinement of the ordering on sequences.

To show that the configurations with DELETE-TO-CONVEX(S) $= 2$ are not well-quasi-ordered, consider the construction in Figure 11.8. We place two points (red) in symmetric positions inside a circle on its horizontal diameter (red dashed line), and another point (yellow) on the intersection of the circle with its vertical diameter. We then form a symmetrical chain of line segments through two points on the circle and one red point inside it, by starting with the segments through the middle yellow point and the two red points, and then repeatedly extending the chain by one more segment at each end. In the figure, this chain is eight segments long. Finally, we add four more "guard" points (blue) on the circle, on either side of the ends of the chain, without creating any additional collinearities.

The construction relies on the fact that the line segments in the chain of segments converge to but never actually reach the horizontal diameter of the circle. At each extension of the chain, the new endpoints are placed on the two arcs of the circle (formed by the previous points placed on the circle) that contain the two endpoints of the horizontal diameter. So no configuration of this type can contain another subconfiguration of the same type, because any subconfiguration either would have more than one chain of three-point lines or its guard points (the points belonging to no three-point lines) would not fall into four separate arcs in the partition of the circle made by the points of the chain. Therefore, the configurations constructed in this way form an infinite antichain with DELETE-TO-CONVEX(S) $= 2$. □

Open Problem 11.25

How many configurations S (as a function of the size of S) have DELETE-TO-CONVEX(S) $< k$?

12 More on Convexity

A puzzle from the Russian mathematical olympiads asks for a proof that, for any convex pentagon of points in a grid, there is another grid point on or inside the smaller convex pentagon formed by the diagonals of the given pentagon (Figure 12.1). Thus, the grid has no empty pentagons.

The existence or nonexistence of empty polygons is not a monotone property, but it suggests the study of other properties, beyond the basic ones studied in Chapter 11, based on what the convex polygons of a configuration contain and not just which points form their vertices.

12.1 Weak Convexity

A configuration S is in *weakly convex position* if none of the points is interior to the convex hull of S: each point either is a vertex of the convex hull (as in configurations that are in convex position) or lies on a convex hull edge. For instance, GRID(2, n) is weakly convex for every n, but GRID(3, 3) is not.

It is tempting to guess that the weakly convex configurations are defined by the property FORBIDDEN(TETRAD), by analogy to the fact that CONVEX = FORBIDDEN(TETRAD, LINE(3)). The three-point line is allowed in weakly convex positions, so it should be dropped from the list of obstacles. However, this guess is not quite right. There is one configuration that is not in weakly convex position but has no tetrad: the *quincunx* (Figure 12.2), a configuration formed from the four points of a convex quadrilateral together with one more point where the diagonals of the quadrilateral cross.[1]

[1] The word "quincunx" usually means four points in a square with a fifth point at its center, but for an early reference to this more general meaning, see Jackson (1821).

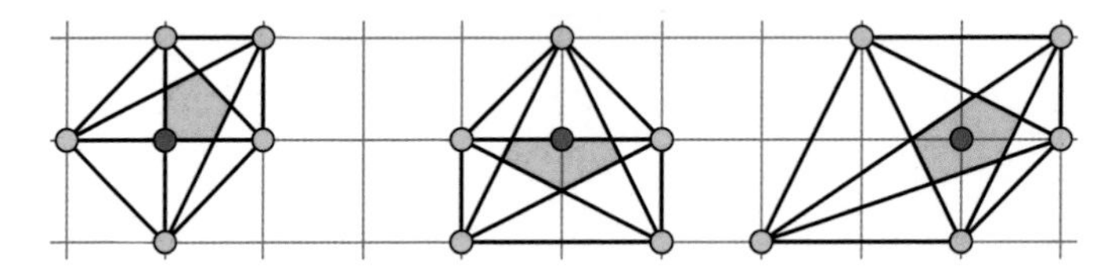

Figure 12.1. Every grid pentagon (yellow) has another grid point (red) on or inside the inner pentagon formed by its diagonals.

The two obstacles LINE(3) and TETRAD for convex position were derived from Carathéodory's theorem, that every point in the convex hull of a set of points belongs to the convex hull of two or three points from the set. The corresponding result for weakly convex position is Steinitz's theorem that every point interior to the convex hull is interior to the convex hull of three or four points from the set.[2] Therefore, the two forbidden patterns for weakly convex position are the tetrad and the quincunx.

Definition 12.1

We define

$$\text{WEAKLY-CONVEX} = \text{FORBIDDEN}(\text{TETRAD}, \text{QUINCUNX})$$

$$\text{WEAK-PARTITION} = \text{PARTITION}(\text{TETRAD}, \text{QUINCUNX})$$

Figure 12.3 depicts examples of these properties and parameters. One could also define in the same way parameters for the largest weakly convex subconfiguration of a given configuration, and for the fewest points to remove to produce a weakly convex subconfiguration. We are not aware of past research on these variants. However, the following result on the largest weakly convex subconfiguration follows from Theorem 11.20, since neither TETRAD nor QUINCUNX is a line or a polygon.

Observation 12.2

$$\text{AVOIDS}(\text{TETRAD}, \text{QUINCUNX}) \doteq \text{SIZE}.$$

Weak convex position and convex position are the same for configurations in general position, as they differ only in allowing the boundary edges of a convex hull to belong to lines of three or more points.

[2] Steinitz (1913).

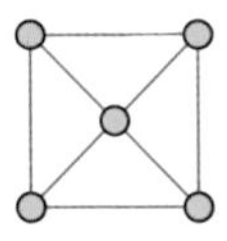

Figure 12.2. The quincunx, an obstacle for weakly convex configurations but not for convex configurations.

12.2 Complexity and Well-Quasi-Ordering

Many of the computational properties of convex sets extend directly to weakly convex sets, using the same methods.

- The property WEAKLY-CONVEX can easily be tested by any convex hull algorithm.
- Theorem 11.15 can be extended to give a polynomial-time algorithm to test whether

$$\text{WEAK-PARTITION}(S) < k,$$

  for any fixed k.
- Like CONVEX-PARTITION, WEAK-PARTITION can be approximated to within a logarithmic approximation ratio by a greedy algorithm that repeatedly finds and removes the largest weakly convex subconfiguration.
- One can ask analogous questions to Open Problem 11.18 on the complexity of approximating WEAK-PARTITION and to Open Problem 11.16 on the fixed-parameter tractability of WEAK-PARTITION.
- And because the obstacles constructed in Theorem 11.11 are in general position, the same construction can be used to prove that there are infinitely many obstacles for WEAK-PARTITION$(S) < 3$.

Unlike convex configurations, there is more than one weakly convex configuration of each size. However, they are still fewer in number than some other classes of configurations.

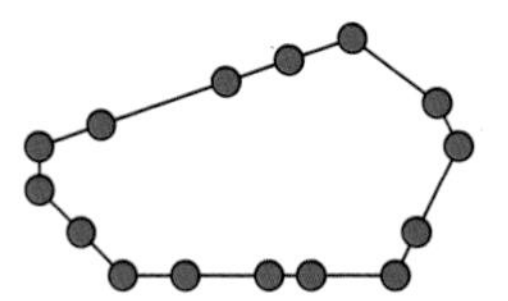

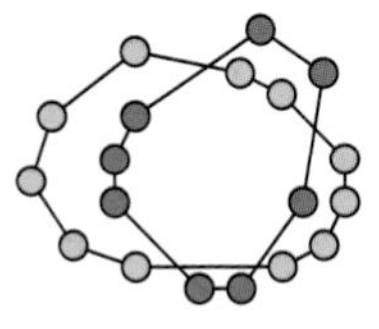

Figure 12.3. A configuration that is in weakly convex position but not convex position (left, red), and another configuration with CONVEX-PARTITION$(S) = 2$ (right, blue and yellow). The union of these two configurations has WEAK-PARTITION$(S) = 3$.

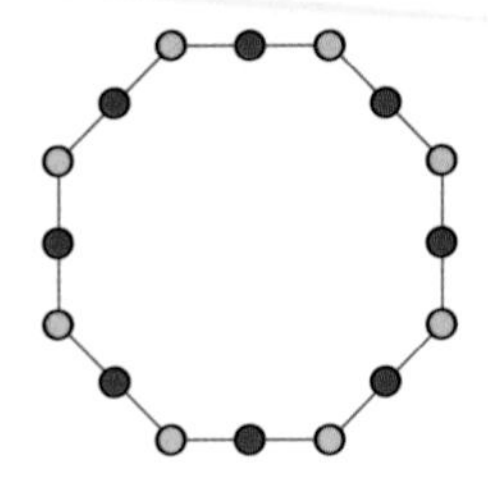

Figure 12.4. Configuration formed from a polygon by subdividing each edge

Theorem 12.3

The number of configurations in weakly convex position on n points is $O(2^n)$. The weakly convex configurations are not well-quasi-ordered.

Proof. Any such configuration can be uniquely described as a cyclic sequence of bits, indicating for each point whether it is a convex hull vertex or whether it lies along a convex hull edge. There are 2^n possible such sequences from any arbitrarily chosen starting point.

For an infinite antichain of weakly convex configurations, consider the configurations formed from POLYGON(n) by adding one point on each polygon edge (Figure 12.4). If a line passes through one of these subdivided polygons but does not contain a polygon edge, it contains only two points of the configuration. Therefore, every proper subconfiguration of one of these subdivided polygons has a convex hull edge with only two points on it. Since no subdivided polygon can be a subconfiguration of any other subdivided polygon, these configurations form an infinite antichain. □

12.3 Crossing Families

Another way to generalize the convex point sets uses *crossing families*, collections of line segments having endpoints in a given configuration such that each pair of line segments in the collection crosses properly.[3] For any n, the configuration POLYGON($2n$) contains a crossing family of n segments, the long diagonals of the polygon. Therefore, the size of the largest crossing family in any configuration S is at least MAX-CONVEX(S). By the happy ending theorem, for configurations S that are in general position, it is $\Omega(\log|S|)$, but Aronov et al. (1994) showed that it is always significantly larger, $\Omega(\sqrt{|S|})$.

[3] Crossing properly means the two line segments have a single intersection point, interior to both segments. Different pairs of segments may have different intersection points.

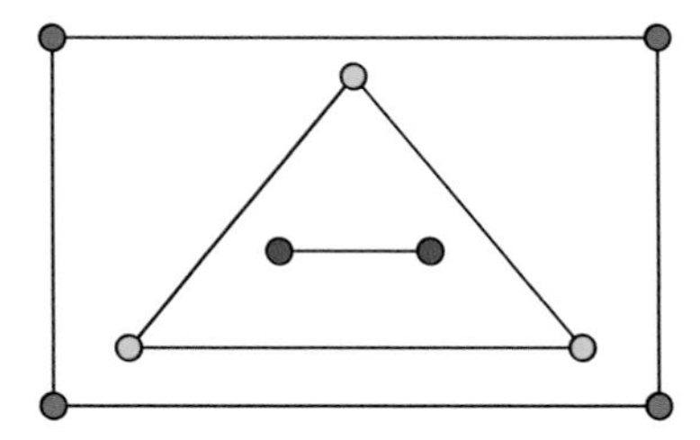

Figure 12.5. The onion layers of a configuration S with $\textsc{onion}(S) = 3$.

Open Problem 12.4

Are there configurations in general position for which the largest crossing family has size smaller than the size of the configuration by a nonconstant factor?

Open Problem 12.5

What is the computational complexity of finding the largest crossing family in a given configuration?

12.4 Onion Layers

The *onion layers* or *convex layers* of a configuration S are a partition of S into nested strictly convex sets. The outermost layer is the convex hull, the second-outer layer is the convex hull of the points that are not in the outermost layer, and so on. Figure 12.5 shows an example. Onion layers have been used to define a notion of depth in robust statistics (related to the Tukey depth of Section 12.7)[4] and in the design of efficient data structures for retrieving points in halfplanes.[5]

Definition 12.6

We define the parameter $\textsc{onion}(S)$ to be the number of layers in the onion layers of S.

[4] Barnett (1976); Eddy (1982). [5] Chazelle et al. (1985).

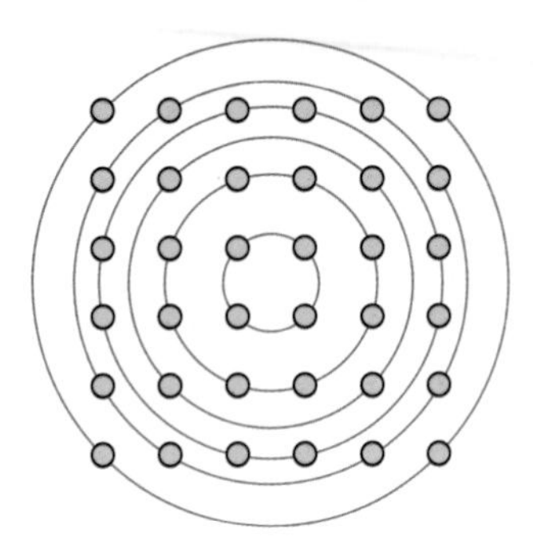

Figure 12.6. The onion layers of GRID(6, 6).

For example, for the configuration S in Figure 12.5, ONION$(S) = 3$. This parameter is not an instance of PARTITION$(C_1, C_2, \ldots)$ because of the constrained location of each layer within the outer layers.

Theorem 12.7 (Har-Peled and Lidický, 2013)

For any positive integer n,

$$\text{ONION}\big(\text{GRID}(n, n)\big) = \Theta(n^{4/3}).$$

A lower bound, that ONION$\big($GRID$(n, n) = \Omega(n^{4/3})$, follows immediately from Observation 11.7, which prevents any of the onion layers from having more than $O(n^{2/3})$ grid points. As Har-Peled and Lidický (2013) show, the actual number of onion layers for the grid is within a constant factor of this lower bound.

Figure 12.6 depicts an example of the onion layers of a square grid, showing in this case that ONION$\big($GRID$(6, 6)\big) = 6$.

Computing the onion layers by repeatedly finding and removing convex hulls takes polynomial time, but is unnecessarily slow. Chazelle (1985) found a faster algorithm.

Theorem 12.8 (Chazelle, 1985)

The onion layers of any configuration S with $|S| = n$ can be found in time $O(n \log n)$.

Another related monotone parameter, the minimum number of subconfigurations in a partition into disjoint convex polygons (meaning that their interiors, and not just their vertex sets, must be disjoint) has also been studied.[6] Like CONVEX-PARTITION, the partition into a minimum number of disjoint convex polygons is NP-hard and has a logarithmic approximation ratio.[7]

Theorem 12.9

ONION *has finite but not polynomial obstacle size.*

Proof. We prove more precisely that the largest obstacle for the property $\text{ONION}(S) < k$ has size $3 \cdot 2^{k-1} - 2$. Each obstacle for this property has k layers $L_1, \ldots L_k$, with one point in its central layer L_k, and with each remaining layer L_i tightly enclosing the next inward layer L_{i+1}, in the sense that no strict subset of L_i also encloses L_{i+1}. For, if not, we could remove a point and still have k layers, so the configuration would not be minimal. On the other hand, every configuration that has these properties is an obstacle. For, removing any point from any layer of the configuration would cause a cascading sequence of changes in which at least one point is removed from each successive layer farther in (by being promoted to another outer layer), eventually completely removing the innermost layer.

In these obstacles, L_{k-1} may have three points, but each remaining layer L_i ($i < k-1$) may have at most twice the number of points of L_{i+1}. For, in order for L_i to tightly enclose L_{i+1}, it is necessary that each consecutive triple of points p, q, r around polygon L_i have a convex hull that includes at least one point of L_{i+1}. If not, q could be removed and L_i would not tightly enclose L_{i+1}. But each point of L_{i+1} can belong to the convex hull of at most two of these consecutive triples, except when L_i is a triangle (as may be the case for L_{k-1}). Therefore, adding the largest possible size for each layer, the total size of the obstacle is at most

$$1 + 3 + 3 \cdot 2 + 3 \cdot 2^2 + \cdots + 3 \cdot 2^{k-2} = 3 \cdot 2^{k-1} - 2.$$

As Figure 12.7 shows, it is possible to achieve this bound by using tightly nested regular polygons of sizes $3 \cdot 2^i$, placed with a common center point and each having a horizontally aligned side. □

Similar parameters have also been defined using partitions into subconfigurations that form either convex polygons or *pseudotriangles* (simple polygons with exactly three convex vertices). However, these parameters are not monotone. Even the property of forming the vertices of a single pseudotriangle is nonmonotone: SAWTOOTH(3) (Figure 3.5) can be connected into a pseudotriangle, but deleting any of its three convex hull vertices produces a nonconvex configuration with four convex hull vertices, something impossible for a pseudotriangle. Additionally, for purposes of approximation, these parameters are not significantly different from the corresponding parameters for

[6] Urabe (1996). [7] Arkin et al. (2003).

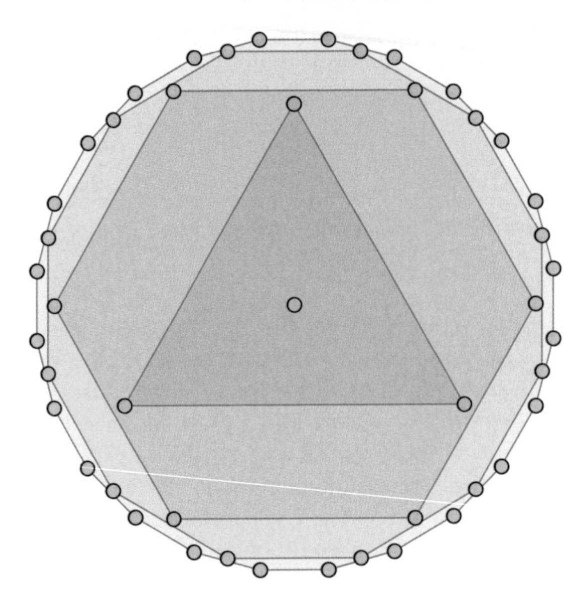

Figure 12.7. Tightly nested regular polygons provide the largest possible obstacles for ONION. Shown is a 46-point obstacle for $\text{ONION}(S) < 5$.

convex polygons. This is because any pseudotriangle can be partitioned into three convex chains, so corresponding convex and convex + pseudotriangle partition numbers are within a factor of three of each other.[8]

12.5 Inequalities

Let's augment the relations between convexity-based parameters that we proved in Section 11.6 with a few more that use weak convexity or onion layers.

Observation 12.10

$$\text{WEAK-PARTITION} \ll \min(\text{CONVEX-PARTITION}, \text{LINE-COVER}).$$

Proof. For every configuration S,

$$\text{WEAK-PARTITION}(S) \leq \text{LINE-COVER}(S)$$

and

$$\text{WEAK-PARTITION}(S) \leq \text{CONVEX-PARTITION}(S),$$

[8] Aichholzer et al. (2007).

because every line or convex set is also weakly convex. For a family of configurations where WEAK-PARTITION is bounded but

$$\min(\text{CONVEX-PARTITION}, \text{LINE-COVER})$$

is unbounded, consider the configurations formed as the union of LINE(n) and POLYGON(n). □

Observation 12.11

CONVEX-PARTITION $\ll$ ONION $\ll$ DELETE-TO-CONVEX.

Proof. CONVEX-PARTITION(S) $\leq$ ONION(S) because the onion layer decomposition is a special kind of convex decomposition. For a family of configurations for which CONVEX-PARTITION is bounded and ONION is unbounded, place two congruent regular n-gons side-by-side. Therefore, the left one of these two relations is true.

If C is the largest convex subconfiguration of any configuration S, and L is an onion layer of S that includes at least one point that is not part of C, then removing L from S reduces DELETE-TO-CONVEX by at least one. Once these onion layers use up all of the points except those in C, the next layer will include all remaining points of C. Therefore, for every configuration S,

$$\text{ONION}(S) \leq \text{DELETE-TO-CONVEX}(S) + 1.$$

The SAWTOOTH configurations have ONION(S) ≤ 2 and DELETE-TO-CONVEX(S) unbounded. □

We can also use onion layers to relate the parameters for partitions into convex position and into weakly convex position subconfigurations.

Observation 12.12

For every configuration S in weakly convex position,

$$\text{ONION}(S) = \left\lceil \frac{\text{ONLINE}(S)}{2} \right\rceil.$$

Proof. For a configuration in weakly convex position, the only lines of three or more points are the ones through the convex hull edges. Each onion layer removes the outermost two points from each such line, until all points on the line have been removed, giving the above formula. □

Corollary 12.13

$$\text{CONVEX-PARTITION} \doteq \text{ONLINE} + \text{WEAK-PARTITION}$$

Proof. If a configuration S has been decomposed into weakly convex subconfigurations, we may separately decompose each subconfiguration into its onion layers, giving a decomposition into at most

$$\left\lceil \frac{\text{ONLINE}(S)}{2} \right\rceil \cdot \text{WEAK-PARTITION}(S)$$

convex subconfigurations. In the other direction, we have already seen that

$$\text{ONLINE} \ll \text{GENERAL-PARTITION} \ll \text{CONVEX-PARTITION}$$

and that

$$\text{WEAK-PARTITION} \ll \text{CONVEX-PARTITION}$$

Thus, CONVEX-PARTITION is bounded below by the two parameters and bounded above by their product. The result follows by Observation 5.19. □

12.6 Polygons Containing Many Points

The problem of finding a k-vertex convex polygon with vertices from S that contains as many other points as possible in S was considered by Eppstein et al. (1992), who showed that for any fixed k it can be solved in time $O(|S|^3)$ by a dynamic programming algorithm similar to the one in Section 11.4. For each choice of k and of whether the number of vertices should be exactly k or at most k, we have a monotone parameter, the number of points of S contained in the optimum polygon (or zero if no such polygon exists). As well as being polynomial-time computable, these parameters have polynomial obstacle size, as an obstacle is a polygon with k or fewer vertices that contains more points than the given parameter value.

Example 12.14

Choose $n > k$ and let $S = \text{SAWTOOTH}(n)$. Then any convex polygon P in S with at most k vertices contains fewer than k additional points of S. Any point that it contains must be near an edge of both P and the convex hull of S, and there can be at most $k - 1$ such edges.

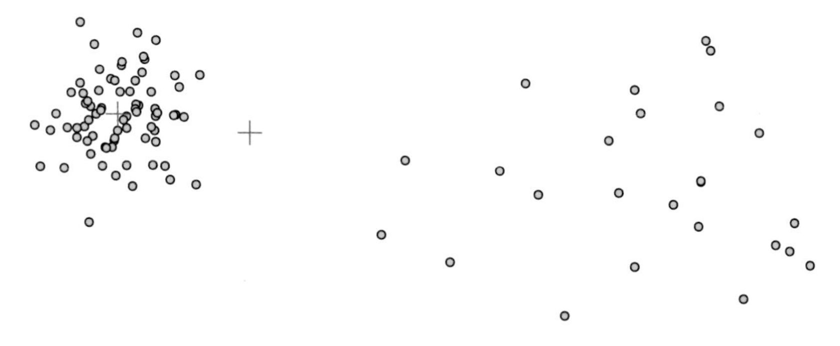

Figure 12.8. 100 points generated from a mixture of two Gaussian distributions. The center of the more heavily populated distribution (a cluster of 75 points) is marked in blue; the centroid of the whole sample is marked in red.

For any k, the sawtooth configurations SAWTOOTH(n) with $n > k$ form an infinite antichain. This example shows that they all have the property that their k-vertex polygons have at most $k-1$ interior points. Therefore, the configurations with this property are not well-quasi-ordered.

For any configuration S, let IN-TRIANGLE(S) be the largest number of points interior to any triangle of S. This is monotone, with its obstacles consisting of triangles with one too many interior points. Pavel Valtr has shown an analogue of the happy ending theorem for this parameter and empty polygons: for every k and n, all sufficiently large general-position configurations with IN-TRIANGLE$(S) < k$ contain an empty convex n-gon.[9]

12.7 Depth and Deep Points

Suppose you are given a sample of data points, which you expect to be clustered around some central point. How should you estimate where that center is? A common answer (but one that we shall see has some problems) is the *centroid* of the points. Computing the centroid is easy: its coordinates are the average of the coordinates of the points. The centroid minimizes the sum of squared distances to the sample points and can be an accurate estimator when the points are symmetrically distributed around the unknown center. But for point sets with many *outliers*, samples far from the center, the centroid can be highly unrepresentative of the data. The outliers can have too big an influence, and pull the centroid away from the main cluster of points. For instance, Figure 12.8 depicts an input consisting of two clusters of points, with the centroid far from either of them. We need some other way of locating the center of the cluster that is more *robust*, more difficult for a small number of outlying points to influence. For one-dimensional points, the *median* provides

[9] Károlyi et al. (2001).

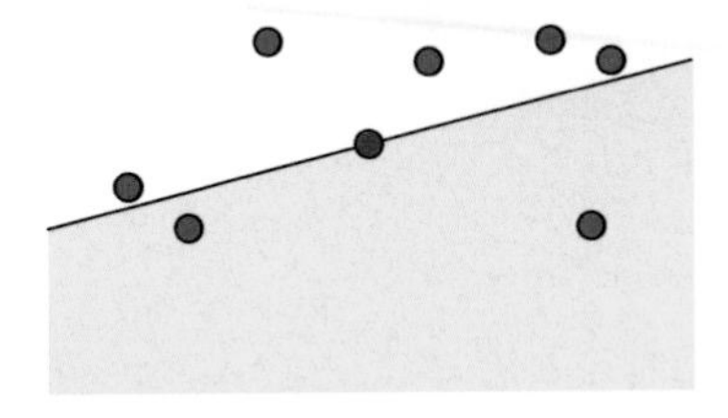

Figure 12.9. The Tukey depth of the red point is determined by finding the closed halfplane through it that contains the fewest points of the sample. In this example, assuming that the red point is itself part of the sample, the yellow halfplane contains three sample points and no other halfplane contains fewer. Therefore, the depth of the red point is 3.

such a robust estimator. If even a bare majority of the points belong to a tight cluster, the median will lie within that cluster, no matter where or how far the other outlying points go. Can we generalize these robustness properties of the median to points in two or more dimensions?

An answer to this question was provided by Tukey (1975). The *Tukey depth* or *data depth* of a point q, not necessarily part of a given collection S of sample points, is the minimum number of points of S in a closed halfplane containing q. That is, among all of the halfplanes containing q, we choose the one with the fewest points of S, and we define the depth of q as the number of points of S in it (Figure 12.9).[10] This can be seen as a "local" analogue of DELETE-TO-CONVEX, as it measures the number of points that must be deleted to cause q to lie outside the convex hull, rather than the number to be deleted to make the whole configuration be in convex position.

It is not difficult to compute the Tukey depth. The halfplane with the fewest points can always be chosen as having q as one of its boundary points. Therefore, when S has n points, the depth of q can be computed in time $O(n \log n)$ by sorting the points of S radially around q and using that sorted order to determine all the distinct subsets of S that can belong to a halfplane through q. An improved algorithm with time $O(n + d \log d)$, where d is the depth of q, is also possible.[11]

For every set S of n points in the plane, there always exists a point q whose depth is at least $n/3$, called a *centerpoint* of S. The deepest point (also called a *Tukey center* or *Tukey median*) is one such point, although there may be others. When the points in S are drawn by adding errors or noise to a single point p, the centerpoint (or a deepest point) of S provides a robust statistical estimator of the location of p, generalizing the one-dimensional median. Because its depth is high, it is insensitive to outliers, points of S whose location has been badly corrupted by noise. Up to $n/3$ of the points may be moved to arbitrary locations, and despite that replacement the centerpoint will remain within the convex hull of the remaining points.[12] It is also possible to construct a centerpoint, or

[10] Tukey (1975). [11] Bremner et al. (2008). [12] Donoho and Gasko (1992).

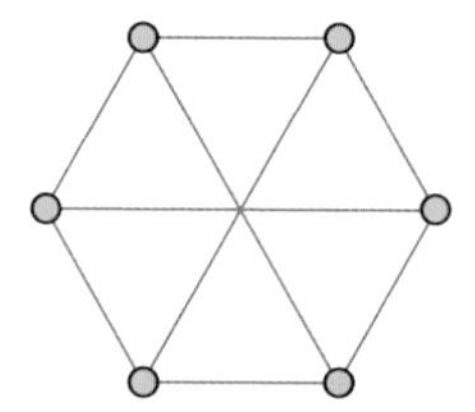
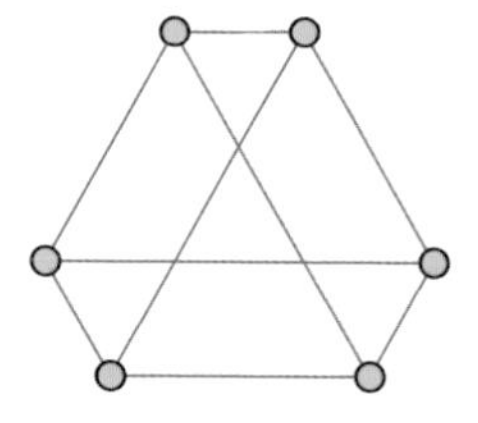

Figure 12.10. Two realizations of POLYGON(6). In the left realization, the three long diagonals meet in a point of depth 3. In the right realization, the three diagonals do not meet in a single point, and all points of the plane have depth at most 2.

a deepest point, efficiently. A centerpoint can be found in linear time,[13] while a deepest point can be found in randomized time $O(n \log n)$.[14]

Unfortunately, the depth of the deepest point of S is not a property of configurations. The reason is that different realizations of the same configuration may have different maximum depths. For instance, a regular hexagon has maximum depth three (at its center). In contrast, a convex hexagon whose three main diagonals do not meet in a single point has depth only two (Figure 12.10). So these two point sets, which form the same configuration, have different maximum depths. Nevertheless, we can define a different monotone parameter by maximizing depth among the points that do belong to S.

Definition 12.15

We define DEPTH(S) to be the maximum data depth, in S, of a point q that also belongs to S.

If we are to represent the whole configuration S by a single point, then the deepest point q is a good choice for a representative: it is the point closest in Tukey depth to the center of q. Because its definition depends only on the radial sorted order of points around q, for a point q in S, it is the same for all realizations of S and gives us a monotone parameter. However, unlike for centerpoints, its depth may be small. For instance, $\text{DEPTH}(\text{POLYGON}(n)) = 1$ because, for each point of POLYGON(n), there is a closed halfplane that contains only that point.

Observation 12.16

DEPTH *has polynomial obstacle size.*

[13] Jadhav and Mukhopadhyay (1994). [14] Chan (2004).

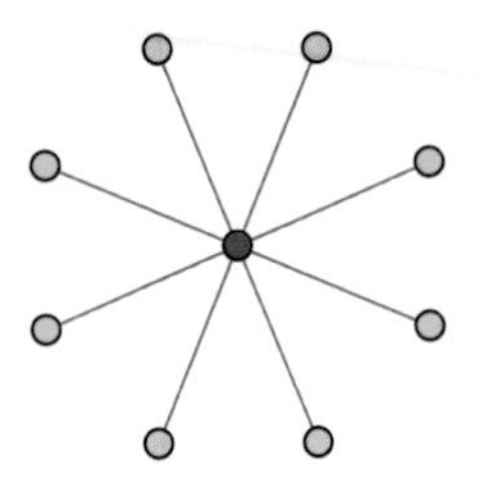

Figure 12.11. Configuration formed from a regular polygon by adding one point at its center

Proof. An obstacle for DEPTH is a configuration S with $\text{DEPTH}(S) \geq k$ but for which all proper subconfigurations have depth less than k. Let q be a point of such a configuration, such that the depth of q is at least k, and let H be a closed halfspace whose boundary passes through q and through at least one other point r. Then H can contain at most $2k-1$ points. For, the k most-clockwise points (including q itself) and the k most-counterclockwise points (again including q) together contain at most $2k-1$ points, and these two subsets of k points prevent any halfplane containing either one of them from having fewer than k points. If any other points belonged to H, they could be removed, without causing any halfplane through q to have fewer than k points. Since S is assumed to be minimal, no other such points can belong to H.

By the same reasoning, the closed halfplane complementary to H also contains $2k-1$ points, including q and r. Together these halfplanes cover the entire plane, so S can have at most $4k-3$ points in total. □

Observation 12.17

$$\text{ONION} \ll \text{DEPTH}.$$

Proof. For any configuration S, $\text{ONION}(S) \leq \text{DEPTH}(S)$, because for each point q in the innermost onion layer, each halfplane containing q contains at least one point in each onion layer. For a family of configurations for which ONION is bounded and DEPTH is unbounded, consider the configurations formed from a regular polygon by adding one point at the center of the polygon (Figure 12.11). □

Observation 12.18

DEPTH *and* DELETE-TO-CONVEX *are incomparable.*

Proof. The configurations formed from a regular polygon by adding one point at the center of the polygon have bounded DELETE-TO-CONVEX but unbounded depth. The SAWTOOTH configurations have bounded DEPTH but unbounded DELETE-TO-CONVEX. □

Observation 12.19

DEPTH *and* LINE-COVER *are incomparable.*

Proof. The configurations GRID$(3, n)$ have bounded LINE-COVER but unbounded DEPTH. The configurations POLYGON(n) have bounded DEPTH but unbounded LINE-COVER. □

Theorem 12.20

For any configuration S of size n, given by the coordinates of a realization, we can compute DEPTH(S) *in time $O(n \log^5 n)$.*

Proof. We apply an algorithm of Matoušek (1991a), which computes a description of the set of points of depth d (as a convex polygon) in time $O(n \log^4 n)$. By checking whether each point of S lies inside or outside this polygon, we can determine whether there exists a point of S whose depth is at least d, in the same amount of time. To find DEPTH(S), we use binary search to find the largest value of d for which such a point exists. □

Open Problem 12.21

Can the exponent of the logarithm in this time bound can be improved? Can DEPTH be computed as efficiently for configurations given only by their order type?

Observation 12.22

The configurations with bounded values of DEPTH *are not well-quasi-ordered.*

Proof. The configurations SAWTOOTH(n) provide an infinite antichain within which

$$\text{DEPTH}\big(\text{SAWTOOTH}(n)\big) = 2. \qquad \square$$

Simplicial depth is another notion of depth in which one counts the number of triangles of sample points surrounding the given point q.[15] Like Tukey depth, the maximum simplicial depth attained anywhere in the plane is not a parameter of configurations, but the maximum attained at one of the given points of a configuration is a monotonic parameter. It can be computed in polynomial time[16] and has obstacles of polynomial size (a point surrounded by k triangles together with the vertices of the triangles).

[15] Liu (1990); Burr et al. (2004).

[16] Khuller and Mitchell (1990); Gil et al. (1992).

13 Integer Realizations

Every general-position configuration can be realized by points with integer coordinates. However, realizing a configuration that has three or more points in a line may require the use of noninteger coordinates. The existence of an integer-coordinate realization is a monotone property, as is the existence of a realization for which all distances are integers.

Definition 13.1

We define the property INTEGER-COORDINATES(S) to be true when S can be realized by points all of whose Cartesian coordinates are integers. We define the property INTEGER-DISTANCES(S) to be true when S can be realized by points all of whose pairwise Euclidean distances are integers.

Another way of defining INTEGER-COORDINATES(S) is that it is the property of being a subconfiguration of GRID(n) for some n. On the other hand, it is not known whether all grids can be realized with integer distances. Figure 13.1 shows an integer-distance realization of GRID$(2, 3)$.

Open Problem 13.2

Which grids can be realized with integer distances?

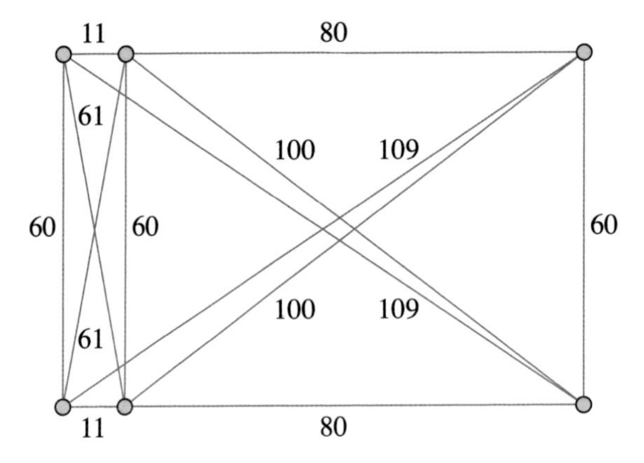

Figure 13.1. An integer-distance realization of GRID(2, 3)

We could equivalently substitute the rational numbers for the integers. Any set of points whose coordinates or distances are rational can be scaled to make them integers, without changing its order type.[1] Klee and Wagon (1991) trace the history of the integer distance problem (in its equivalent formulation with rational numbers) to the seventh-century Indian mathematician Brahmagupta, who already asked about rational-distance realizations of POLYGON(4).

13.1 The Perles Configuration

The realizability of configurations using rational coordinates was considered by Micha Perles in the 1960s. Perles found an unrealizable configuration now called the Perles configuration, consisting of nine points at the corners and center of a regular pentagram (Figure 13.2).

The properties of this configuration depend on the projective transformations of the plane and on a number defined from quadruples of collinear points called their *cross-ratio*. A projective transformation is a transformation of the plane that preserves collinearities among all of the (infinitely many) triples of points in the plane. The cross-ratio of four distinct points A, B, C, and D appearing in that order along a line is defined as

$$\frac{\text{DIST}(A, C) \cdot \text{DIST}(B, D)}{\text{DIST}(A, D) \cdot \text{DIST}(B, C)},$$

where DIST denotes the Euclidean distance. A projective transformation might reverse the ordering of some pairs of points, and then their distances are negated in the formula.

[1] This argument only works for finite point sets, because for infinite point sets with rational distances there may be no scaling factor that makes all distances integers. According to the Erdős–Anning theorem (Anning and Erdős, 1945), the only infinite point sets with integer distances are subsets of the points on a single line.

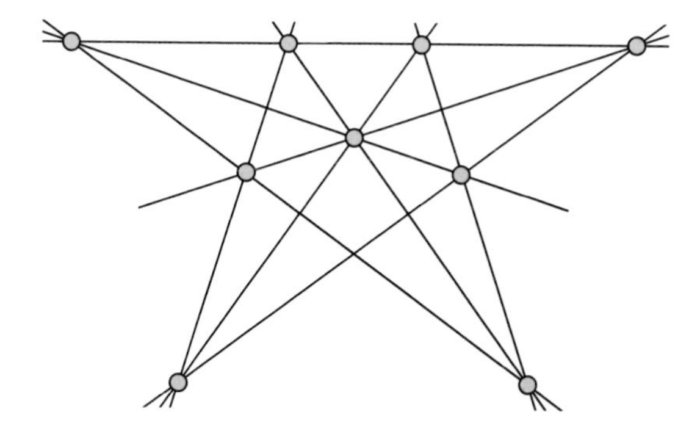

Figure 13.2. The Perles configuration

The key properties needed to understand the Perles configuration are the following.[2]

- Projective transformations always preserve the cross-ratio of transformed quadruples of collinear points.
- The cross-ratio of points with integer coordinates or integer distances is a rational number. For integer distances this follows directly from the formula for the cross-ratio. For integer coordinates, each line through two or more points has a rational number as its slope, and if we represent this number in lowest terms as a/b, then the distances between pairs of points along this line are (by the Pythagorean theorem) all integer multiples of $\sqrt{a^2 + b^2}$. This number is not always rational, but it appears in all four distances in the cross-ratio formula and is therefore canceled out.
- The Perles configuration is projectively rigid, meaning that any two realizations of this configuration can be mapped into each other by projective transformations (although these transformations may permute the points of the configuration).
- Because of its rigidity and from its construction (in the drawing shown in Figure 13.2) from a regular pentagon, in any realization of the Perles configuration its four collinear points have a cross-ratio equal to

$$\frac{3+\sqrt{5}}{2} \text{ or } \frac{3-\sqrt{5}}{2},$$

  irrational numbers closely related to the golden ratio.

Therefore, this configuration has no realization with integer coordinates, and no realization with integer distances.

[2] Grünbaum (2003); Ziegler (2008).

Open Problem 13.3 (Grünbaum, 2003)

Is the Perles configuration the smallest possible obstacle for INTEGER-COORDINATES?

More generally, very little is known about the obstacles for these problems.

Open Problem 13.4

Are there finitely many obstacles for the properties of having realizations with integer coordinates or with integer distances? What is the computational complexity of testing these properties? Do they have sublinear property testing algorithms?

The existence of a realization with integer coordinates, for a given configuration, can be expressed by a Diophantine equation, a polynomial equation with integer coefficients that must be solved by assigning integer values to its variables. However, there cannot exist an algorithm to solve all Diophantine equations.[3] This fact raises the possibility that testing INTEGER-COORDINATES might also be undecidable.

13.2 Polygons

As Euler (1862) already proved, one can find arbitrarily large general-position but cocircular configurations of points S with all distances rational. Harborth (1998) gives two different constructions.

- Let θ be one of the non-right angles in an integer-sided right triangle, for instance
$$\theta = \sin^{-1}\frac{3}{5}$$
from a $(3,4,5)$ right triangle. Then the points $(\cos i\theta, \sin i\theta)$ for $i = 0, 1, 2, \ldots$ formed by rotating a unit vector by integer multiples of θ produce a dense subset of the unit circle with all distances rational. This construction is shown in Figure 13.3, left.
- Start with any triangle with rational sides, inscribed in a circle, and generate more points by repeatedly reflecting the triangle across the perpendicular bisectors of its sides. Each reflection generates another point on the circle (except when an isosceles triangle is reflected perpendicularly to its base).

[3] Matiyasevich (1993).

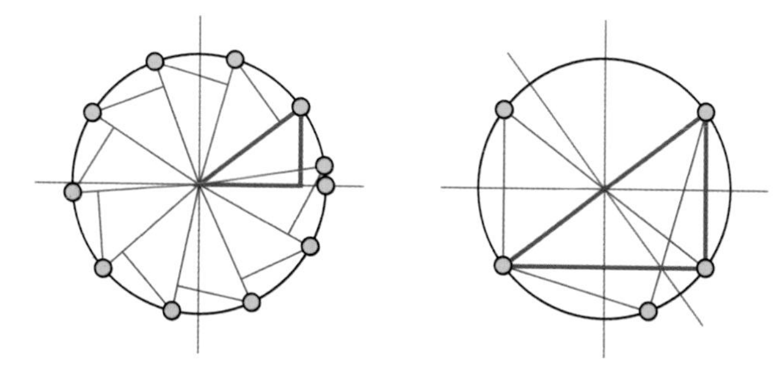

Figure 13.3. Two constructions of Harborth (1998) for sets of cocircular points with all distances rational, realizing any POLYGON(n) configuration: Left, rotation by the angle of a Pythagorean triangle. Right, reflection of an integer-sided triangle across its perpendicular bisectors. Both constructions are shown with the $(3, 4, 5)$ right triangle.

The integer sides of the triangle and its reflection together include five of the six distances between their four points. Ptolemy's theorem that the product of diagonals of an inscribed quadrilateral equals the sum of products of opposite sides then implies that the sixth distance is also rational. Repeated application of Ptolemy's theorem shows the same for all of the distances among the generated points. This construction is shown in Figure 13.3, right.

These constructions lead to arbitrarily many (or infinitely many) points on a circle with all distances rational. Therefore, every convex order type POLYGON(n) can be realized with rational or integer distances.

Open Problem 13.5

Can every weakly convex configuration be realized with all distances rational? If not, what are the minimal weakly convex configurations that cannot be realized with all distances rational?

13.3 Points on Two Lines

Suppose that we have two sets of integers $X = \{x_1, x_2, \ldots\}$ and $Y = \{y_1, y_2, \ldots\}$ such that each two numbers (x_i, y_j) drawn from these two sets form the two short sides (legs) of a integer-sided right triangle. By the Pythagorean theorem, these numbers obey the equations $x_i^2 + y_j^2 = z_{i,j}^2$ for integers $z_{i,j}$, the long sides (hypotenuses) of these triangles. We can use these numbers to generate $2|X| + 1$ points $(0, 0)$ and $(\pm x_i, 0)$ on the x-axis of the plane, and $2|Y| + 1$ points $(0, 0)$ and $(0, \pm y_i)$ on the y-axis, all at integer distances from each other. The distances between points on the same axis are the integer differences of their coordinates, and the distances between points that are not the same axis are all

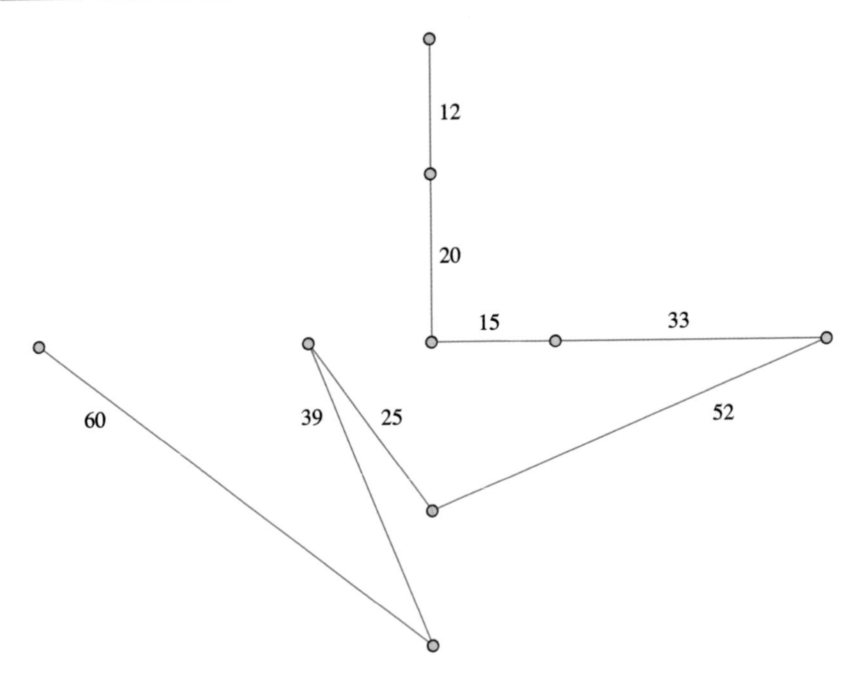

Figure 13.4. An integer-distance set of nine points, five on each of two perpendicular lines, derived from the two sets of Pythagorean triangle legs $X = \{15, 48\}$ and $Y = \{20, 36\}$.

hypotenuses of Pythagorean triangles. Figure 13.4 shows an example with two integers in each set: $X = \{15, 48\}$ and $Y = \{20, 36\}$. Correspondingly, the figure has five points on each axis, including the point at the origin where the two axes cross.

There exist sets X and Y of Pythagorean legs like this with $|X| = 2$ and with Y arbitrarily large.[4] These sets can be used to generate integer-distance configurations on two lines, with five points on one line and arbitrarily many on the other. There also exist sets of Pythagorean legs with $|X| = 3$ and $|Y| = 4$. These can be used to generate integer-distance configurations on two lines with seven points on one line and nine on the other.[5]

Open Problem 13.6

Do there exist sets X and Y of integers such that every $x \in X$ and $y \in Y$ form the legs of a Pythagorean right triangle, with $\min(|X|, |Y|) \geq 3$ and $|X| + |Y| > 7$? If so, how large can these sets be?

[4] Peeples (1954). If we use rational numbers in place of integers, Y can be made infinitely large.

[5] Lagrange and Leech (1986).

13.4 Non-Cocircularity

The large integer-distance sets that we have constructed so far all have many points on a single line or circle.

Open Problem 13.7

Do there exist arbitrarily large sets of points in which the distances are all integers, no three points are on a line, and no four points are on a circle?

The largest such sets that have been found have only seven points.[6]

Cocircularity (the property of a set of points that all lie on a single circle) is not something we can consider directly when we study configurations instead of point sets. This is because two different realizations of the same configuration can have different subsets of cocircular points. However, to some extent, we can use convexity as a stand-in for it. In particular, as the following observation shows, configurations that cannot be partitioned into a small number of convex subconfigurations always have large non-cocircular subsets in any of their realizations (possibly different ones in different realizations).

Observation 13.8

For every k, and every configuration S for which

$$\textsc{convex-partition}(S) \geq k,$$

every realization of S contains $\Omega(k^{1/3})$ points no four of which lie on a circle.

Proof. Given an arbitrary set of points S let C be a subset of S that is maximal with respect to the property of having no four cocircular points. Then every point of S must belong to one of the $O(|C|^3)$ circles through triples of points in C (as shown in Figure 13.5), or else C would not be maximal. Therefore, $\textsc{convex-partition}(S) = O(|C|^3)$. Equivalently,

$$|C| = \Omega(\textsc{convex-partition}(S)^{1/3}). \qquad \square$$

Open Problem 13.9

For configurations S in general position for which $\textsc{integer-distances}(S)$ is true, is $\textsc{convex-partition}(S)$ bounded by a constant?

[6] Kleber (2008); Kreisel and Kurz (2008).

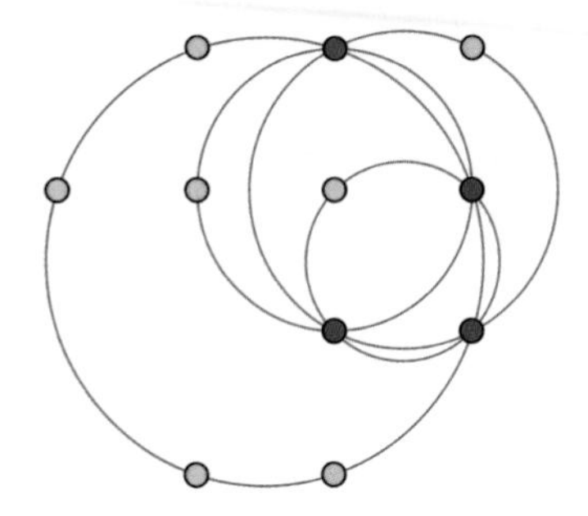

Figure 13.5. For a subset of a set of points that is maximal for the property of having no four cocircular points, such as the red subset of points in the figure, the remaining (yellow) points must all lie on the circles through triples of points in the subset.

By Observation 13.8, a negative answer to Open Problem 13.9 would imply that there exist arbitrarily large sets of points with no three in a line, no four cocircular, and all distances integer. This would be a big extension on the known sets of this type, whose size is only seven.[7] On the other hand, a positive answer to Open Problem 13.9 would impose a significant amount of structure on the integer-distance sets.

13.5 Harborth, Erdős, and Ulam

Harborth's conjecture, an open problem in graph drawing, concerns planar graph drawings with integer edge lengths. The difference between this edge length requirement and the requirements of INTEGER-DISTANCES is that we do not require the other distances, between pairs of vertices that are not adjacent, to be integers. Figure 13.6 gives an example of a graph drawn in this way. This particular drawing can also be realized with integer coordinates, but that condition is not part of the conjecture.

Open Problem 13.10 (Harborth's conjecture (Kemnitz and Harborth, 2001))

Can every planar graph be drawn with straight-line-segment edges without crossings in such a way that each edge has integer length?

[7] Kleber (2008); Kreisel and Kurz (2008).

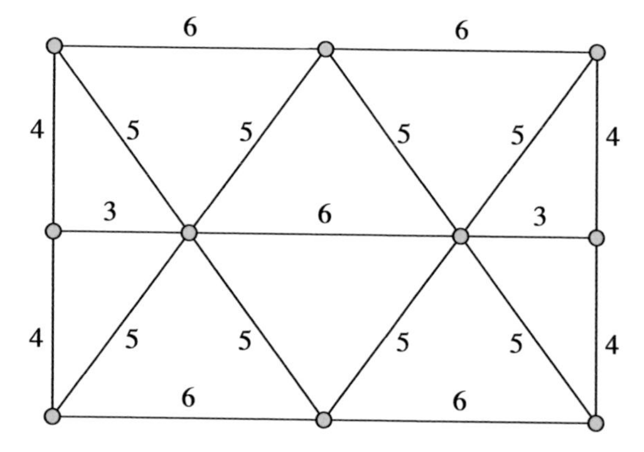

Figure 13.6. A planar graph, drawn with integer edge lengths.

Another open problem, the *Erdős–Ulam conjecture,* states that there is no dense set of points in the plane with rational distances.[8]

Open Problem 13.11 (the Erdős–Ulam conjecture)

Does every dense subset of the plane include two points whose distance is irrational?

An answer to the following question on integer distances would resolve one or the other of these conjectures.

Open Problem 13.12

Does INTEGER-DISTANCES have at least one obstacle in general position?

If Open Problem 13.12 has a positive answer, and there is a general-position obstacle, then (for any set of points S realizing that obstacle) any rational-distance point set would necessarily avoid a neighborhood of one of the points in S. Therefore, there could exist no dense rational-distance point set, and the Erdős–Ulam conjecture would be true. In particular the Erdős–Ulam conjecture would follow from a positive answer to Open Problem 13.9, because then an

[8] Solymosi and de Zeeuw (2010).

obstacle to INTEGER-DISTANCES(S) could be found as a subconfiguration of any general-position configuration with a large enough value of CONVEX-PARTITION.

On the other hand, every planar graph can be realized as a straight-line drawing with vertices in general position.[9] If Open Problem 13.12 has a negative answer, meaning that there is no general-position obstacle to INTEGER-DISTANCES(S), then the vertex positions of any such drawing could be realized with integer distances between all vertices. In this case, Harborth's conjecture would be true.

[9] The theorem that all planar graphs have straight-line drawings is known as Fáry's theorem, but it was independently discovered by Wagner (1936) and Stein (1951) as well as by Fáry (1948).

14 The Stretched Geometry of Permutations

Two rays at right angles to each other, one downward and one horizontal (either left or right) from the same starting point, form a polygonal curve. These curves look like the ceiling symbols $\lceil x \rceil$ used to denote rounding a number x up to the nearest integer, so we call them *ceilings*. Ceilings behave in many ways like lines (Figure 14.1). For instance, if two ceilings cross each other without overlapping, they do so only once, like lines. And any two points in the plane that do not have the same x-coordinates or y-coordinates as each other are connected by a unique ceiling, just as they are connected by a unique line.

The analogy between ceilings and lines can be made precise through a transformation of point sets into "stretched configurations" whose lines correspond to the ceilings of the original point set. The construction replaces the y-coordinates of the points with "large enough" numbers having the same ordering. Configurations constructed in this way have been used for several problems in discrete geometry and graph drawing.[1] They play a key role in the construction of universal configurations that we describe in Chapter 16.

14.1 The Stretching Transformation

Let S be a set of points in the plane – not a configuration, but a set of points, as we need its coordinates. Suppose also that no two points in S have the same x-coordinate as each other, nor the same y-coordinate as each other. Then we can describe any ceiling by the position of its vertical ray in the sorted

[1] Middendorf and Pfeiffer (1992); Bukh et al. (2011); Bannister et al. (2014); Fulek and Tóth (2015).

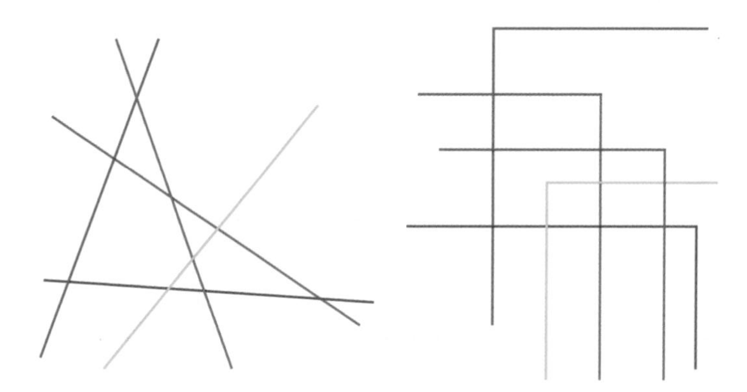

Figure 14.1. An arrangement of lines (left) and an arrangement of ceilings (right) that cross each other in the same pattern. Lines with positive slope correspond to ceilings that extend right, and lines with negative slope correspond to ceilings that extend left.

sequence of x-coordinates, the position of its horizontal ray in the sorted sequence of y-coordinates, and whether the horizontal ray extends left or right. Two ceilings with the same description pass through the same points (if any) of S, have the same points on their left side, and have the same points on their right side. Because this description depends only on the coordinate orderings of S, and not on the numerical values of the coordinates, transformations of S that preserve these orderings cannot change the way that ceilings interact with S.

We will use this invariance under order-preserving transformations to construct a configuration from S, by transforming it in an order-preserving way to make each line of the plane cross the transformed set in an equivalent way to a ceiling and vice versa. After this transformation, we can describe properties of a ceiling (something not directly accessible in a configuration) by the properties of the corresponding line.

Definition 14.1

As above, let S be a set of points no two of which have the same coordinates. We define $\text{STRETCH}(S)$ to be the configuration of the set of points S', constructed from S as follows: for each point (x, y) of S, in order from smaller to larger y-coordinates, place a corresponding point (x, y') in S'. The x-coordinate of this point is the same as in S, but the y-coordinate may differ. The new coordinate y' may be chosen as any number large enough to make the new point (x, y') lie above all previously placed points in S', and above all lines through pairs of previously placed points.

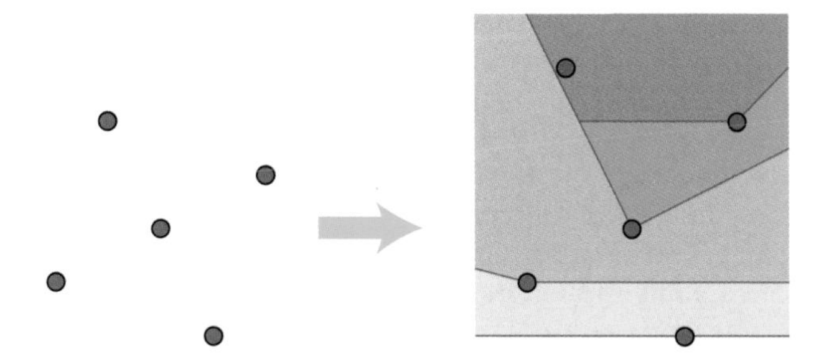

Figure 14.2. A set S of points (left) and STRETCH(S) (right). The red contours show the regions where each point of STRETCH(S) can be placed, so that it is high enough to be above all lines through lower points. The bottom three points keep their original y-coordinates, but the top two have been raised relative to their original positions.

It may not be clear from this definition that the resulting configuration does not depend on the exact choice of how high to make each point, but we will prove this below. Figure 14.2 shows an example of this transformation. In this example, the bottom three points of S' can keep their previous y-coordinates, because they were already above all lower points and above all lines through pairs of lower points. However, the top two points need to be raised so that they move to positions above some lines through lower points. This transformation does not always preserve the order type of S. For instance, some points that were collinear before the transformation are not collinear after it. However, it preserves the coordinate ordering of the points, and therefore it does preserve the interactions of S with all ceilings.

Observation 14.2

Every configuration STRETCH(S) *is in general position.*

Proof. We cannot create three collinear points, because the uppermost of the three would be placed above the line through the other two. □

Observation 14.3

Let p, q, and r be any three points of a set S of points (with no two equal coordinates) and let p', q', and r' be the corresponding three points in STRETCH(S), *with p being the topmost of the three points. Then the orientation of (p', q', r') is $+1$ if q has a greater x-coordinate than r, and -1 otherwise.*

Proof. Because p' is placed above the line through q' and r', the cyclic ordering of the other points as seen from p' (which is what defines the orientation) will always be consistent with their left-to-right ordering by x-coordinates. □

Because the orientations within any configuration STRETCH(S) depend only on the coordinate orderings of the points in S, the choice of how high to place each point within the construction for STRETCH(S) does not affect the result: any two choices for how to assign new y-coordinates to the points of S', consistent with the rule that each point must be placed above all lower points and all lines through pairs of lower points, lead to the same configuration.

This description of the orientations of triples of points is also what you would get by defining the orientation according to whether one point of a triple is on one side or the other of the ceiling through the other two points. Therefore, the partition of STRETCH(S) made by the line through two points is the same as the partition of S made by the ceiling through the corresponding two points. But we have already seen in Corollary 3.13 that, even when a line does not pass through two points of a configuration, it makes the same partition of the points as a small perturbation of another line that does pass through two points. Whatever this perturbation is, we can make the same perturbation to the ceiling. So all lines that pass through STRETCH(S), regardless of whether they are lines through two points of S, are combinatorially equivalent to ceilings and vice versa.

The stretching transformation also interacts well with subconfigurations.

Observation 14.4

If S and T are finite sets of points with no two equal coordinates, and $S \subset T$, then STRETCH(S) *is a subconfiguration of* STRETCH(T).

Proof. Let T' be constructed from T as in the definition of STRETCH(T), and let S' be the subset of T' corresponding to the subset S of T. Then the assignment of y-coordinates to points in S' meets the requirements of the STRETCH construction as well. □

We can use this construction to define a monotone property of configurations.

Definition 14.5

We define the property STRETCHED of a configuration to be true when the configuration can be generated by the STRETCH construction, and false otherwise.

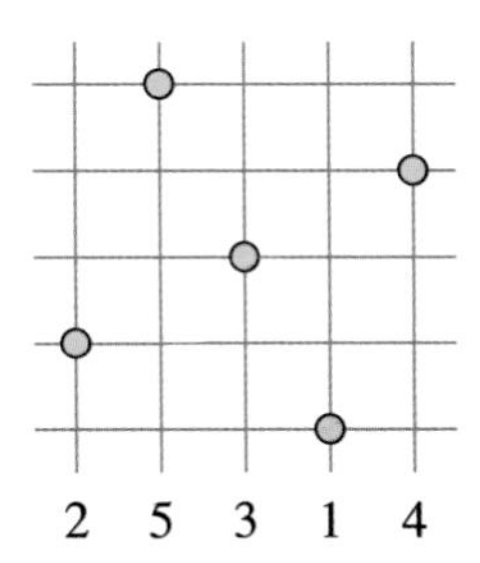

Figure 14.3. The point set PLOT(25314), generated from the permutation 25314. Its stretched transformation is the one shown in Figure 14.2.

14.2 Permutations and Patterns

We have seen that the order type of a configuration STRETCH(S) depends only on the horizontal and vertical ordering of the points in S. The language of *permutations* provides a natural way of describing these orderings.

Definition 14.6

We define a permutation as a sequence of the numbers from 1 to n, for some positive integer n, such that each number appears exactly once in the sequence. When $n < 10$, it is convenient to write these sequences by concatenating the digits from 1 to n in the sequence order. For instance, the six permutations of the numbers from 1 to 3 can be written in this way as 123, 132, 213, 231, 312, and 321. If π is a permutation,[2] consisting of the sequence of numbers $(y_1, y_2, y_3, \ldots, y_n)$, then we define PLOT($\pi$) to be the set of points (i, y_i) obtained by using the sequence positions as x-coordinates and the sequence values as y-coordinates.

Figure 14.3 shows an example of this construction.

Observation 14.7

Every STRETCHED *configuration is of the form*

$$\text{STRETCH}\big(\text{PLOT}(\pi)\big)$$

for at least one permutation π.

[2] It is standard to write Greek letters, and especially the letter π, for permutations. This notation is unrelated to the other meaning of π involving the measurement of circles.

Proof. If a configuration C is of the form STRETCH(S) for a set of points S, we may define a permutation π in which the ith permutation value is the position, in the vertical ordering of S, of the ith point in the horizontal ordering of S. Because π is defined in such a way as to make S and PLOT(π) have the same horizontal and vertical orderings, and because the order type of the stretched configurations of these point sets depends only on these orderings (Observation 14.3), it follows that STRETCH(S) = STRETCH(PLOT(π)). □

We must be careful, however: two different permutations may have the same stretched plots. For instance, all six permutations on three elements generate the same configuration, POLYGON(3). More generally, whenever two permutations on the same n numbers differ only on the position of their maximum value n, and have the same ordering on the remaining values, they will generate the same stretched configurations. Therefore, for every permutation on n numbers there will be at least $n-1$ other permutations that all have the same stretched configuration.

Just as subconfigurations provide a natural definition of a smaller configuration within a larger one, the notion of a *permutation pattern* provides a corresponding concept for permutations. The study of permutation patterns is deep, far-ranging, and largely beyond the scope of this work; see Kitaev (2011) for more on this subject. However, we can at least provide the basic definitions, framed in terms of our geometric view of permutations.

Definition 14.8

If S and T are permutations, then S is a *pattern* of T if there is a subset of PLOT(T) whose points have the same horizontal and vertical orderings as the points of PLOT(S). This subset of PLOT(T) may be called an *instance* of S in T, when that will not cause confusion with our other definition of instances of subconfigurations in configurations.

For example, the leftmost three points of PLOT(25313) (the point set shown in Figure 14.3) form an instance of the pattern 132, because they have the same horizontal and vertical orderings as the three points of PLOT(132).

Observation 14.9

For every permutation π, and every instance C of a subconfiguration of STRETCH(PLOT(π)), *C can also be seen as an instance of a pattern of π.*

As we will see in Theorem 14.26, this correspondence between subconfigurations and patterns will allow us to test for the existence of arbitrary subconfigurations in stretched configurations, much more easily than we can in arbitrary configurations.

Just as we have defined classes of configurations in terms of the subconfigurations that they avoid, researchers in the theory of permutation patterns have defined classes of permutations in terms of the patterns that they avoid. In this context, for any permutation π, the *π-avoiding permutations* are the permutations that do not have π as a pattern. We will return to this concept in Chapter 16, when we use stretched plots of 213-avoiding permutations to draw planar graphs.

14.3 Stair-Convexity

Convexity may be defined for ceilings in the same way that it is for lines, and (because of the equivalence between ceilings and lines in stretched point sets) it behaves in much the same way. In order to define convexity, we first need a concept of a line segment for ceilings.

Definition 14.10

For any two points p and q in the plane, the *ceiling segment* from p to q is a curve with p and q as endpoints. If p and q have an equal coordinate, it is the line segment from p to q, and otherwise it is the portion of the ceiling through p and q that lies between p and q.

Definition 14.11 (Bukh et al., 2011)

A subset of the plane is *stair-convex* if it contains the ceiling segment between every two of its points. The *stair-convex hull* of a set S of points is the intersection of all stair-convex sets that contain S.

We can describe the stair-convex hull of a finite point set S more explicitly. By the equivalence between ceilings and lines of the stretched configuration, this hull has the same cyclic sequence of vertices as the convex hull of STRETCH(S), but connected by ceiling segments instead of line segments. There can only be one ceiling segment from the leftmost vertex to the top vertex and from the top vertex to the rightmost vertex (if these pairs of vertices are distinct). This is because a concatenation of more than one ceiling segment in this direction would not be stair-convex. However, the parts of the stair-convex hull between the leftmost vertex and the bottom vertex, and between the bottom vertex and the rightmost vertex, can be monotone paths that alternate between horizontal and vertical segments arbitrarily many times. Near the bottom vertex itself, the

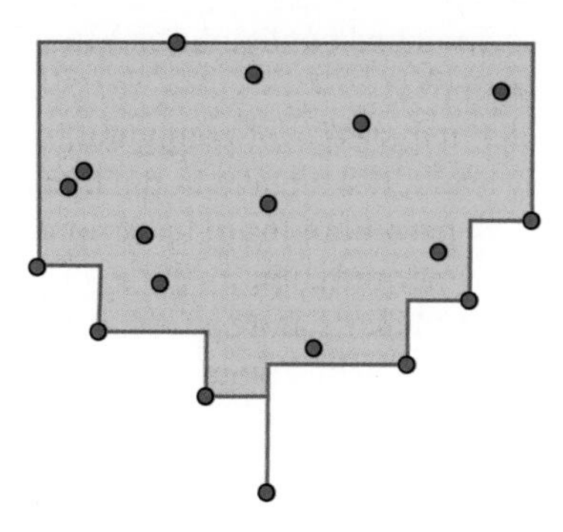

Figure 14.4. The stair-convex hull of a set of points.

stair-convex hull can degenerate to a single vertical line segment, shared by the staircase paths from the bottom vertex to the leftmost and rightmost vertices. Figure 14.4 depicts an example. It is also possible for the leftmost or rightmost vertex to coincide with the top vertex. If it does, the staircase hull can include a degenerate horizontal line segment.

Based on this description, we can characterize precisely the sets of points S that become convex position configurations when they are stretched.

Observation 14.12

Define a sequence of numbers to be downward unimodal *if it can be cut into two monotonic subsequences: it strictly decreases from its first element to its minimum element, and then strictly increases from the minimum element to its last element.*[3] *Then* STRETCH(S) *is in convex position if and only if the sequence of y-coordinates of its points, in left-to-right order with the maximum y-coordinate omitted, is downward unimodal.*

This leads to a significant strengthening of the happy ending theorem (Chapte 1) for these configurations.

Theorem 14.13 (The Erdős–Szekeres theorem (Erdős and Szekeres, 1935))

In any sequence of n values from an ordered set, we can find a subsequence of $\Omega(\sqrt{n})$ values that are all consistently ordered: they are strictly increasing, strictly decreasing, or all equal.

[3] The more usual definition of a unimodal sequence is one that increases to its maximum and then decreases after that element, but we need the reversed concept.

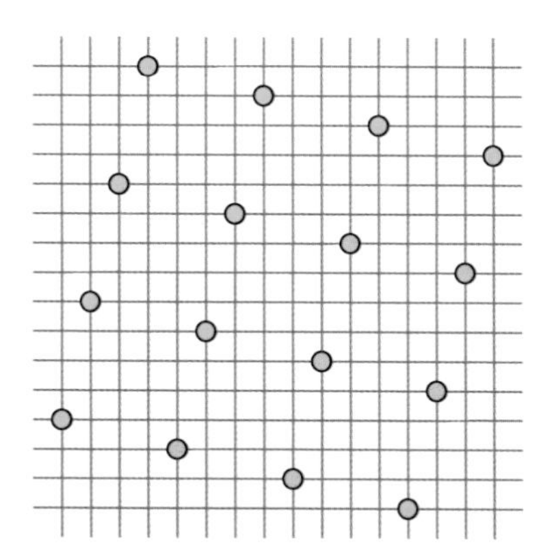

Figure 14.5. A tilted grid, realized as the plot of a permutation. Stretching sets of points of this form produces configurations in which the largest convex subconfiguration has size $O(\sqrt{|S|})$.

Corollary 14.14

Every STRETCHED *configuration S has a convex subconfiguration of size $\Omega(\sqrt{|S|})$.*

The tilted grid of Figure 14.5 shows that this bound is tight, up to constant factors, and provides a family of examples of stretched configurations for which CONVEX-PARTITION is also $\Omega(\sqrt{|S|})$. Because CONVEX-PARTITION can be large for these stretched configurations, they could potentially provide examples that could help resolve Open Problem 13.9, on the existence of integer-distance configurations with large values of CONVEX-PARTITION. This raises the following question.

Open Problem 14.15

Can every STRETCHED configuration be realized with integer distances?

A positive answer to Open Problem 14.15 would also give a positive answer to Open Problem 13.9 and (via the construction of universal configurations by stretching in Theorem 16.7) to Harborth's conjecture (Open Problem 13.10). On the other hand, a negative answer to Open Problem 14.15 would imply that there exists a general-position obstacle to INTEGER-DISTANCES (Open Problem 13.12) and therefore that there is no dense rational-distance subset of the plane (the Erdős–Ulam conjecture, Open Problem 13.11).

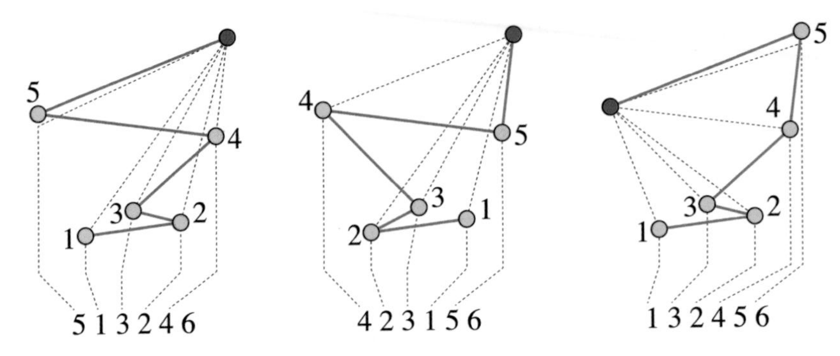

Figure 14.6. Three stretched paths in the same configuration (starting in each case from the red point), and the numbering and cyclic ordering used by UNSTRETCH to construct a permutation from each path.

14.4 Characterization

We can characterize STRETCHED configurations combinatorially, by describing their vertically sorted sequences of vertices in terms of intrinsic features of the configuration rather than in terms of their coordinates. To do so, we need the following definition. This characterization will be useful in describing algorithms to recognize the stretched configurations.

Definition 14.16

Let S be a configuration in general position. Then a *stretched path* in S is an arrangement of the points of S into a sequence such that:

- Every point appears exactly once in the sequence.
- The first point of the sequence belongs to the convex hull of S.
- For every two consecutive points p and q in the sequence, there do not exist two points r and s later in the sequence for which line rs separates p from q.

Figure 14.6 shows three different stretched paths in the same configuration.

Observation 14.17

If $S = \text{STRETCH}(T)$ for a set of points T, then the top-down vertical ordering of T (in sorted order by y-coordinate, from largest to smallest y-coordinate) is a stretched path. Thus, every stretched configuration has a stretched path.

Proof. Every point appears exactly once in this sorted order, and the topmost point necessarily forms a vertex of the convex hull. Each two consecutive points both lie above all lines through later pairs of points in the sequence (by the definition of STRETCH) so no such line can separate these two points. □

We can transform stretched paths into permutations (not necessarily equal to the permutation from which a stretched configuration might have been constructed).

Definition 14.18

Let P be a stretched path on n points, beginning with a point p_0. Then we define a permutation UNSTRETCH(P) on the numbers from 1 to n, as follows. Number the remaining points of P as $p_1, p_2, \ldots, p_{n-1}$, in order along the path with p_1 farthest from p_0 and p_{n-1} adjacent to p_0. Then, sort these points in counterclockwise order around p_0, starting from the counterclockwise neighbor of p_0 along the convex hull and ending at the clockwise neighbor. Write down the sequence of indexes i of the points i, in this sorted order, and complete the permutation by adding the number n at its end.

The numbering of the points and their cyclic order around the starting vertex of the path, used by this construction for UNSTRETCH, are illustrated for three different stretched paths in Figure 14.6.

Lemma 14.19

If P is a stretched path in any configuration S, then

$$S = \text{STRETCH}\Big(\text{PLOT}\big(\text{UNSTRETCH}(P)\big)\Big)$$

Proof. Let p_0 be the first point of the path. Consider the orientation of any triple of points (p, q, r) in S, where we can assume by permuting the three points that p is earlier than q and r in the path P. Then, because line qr does not separate any consecutive pair of points in P that are earlier than q, it follows that (p, q, r) has the same orientation as every triple (p', q, r) where p' is earlier than p in the path, and in particular that (p, q, r) has the same orientation as (p_0, q, r). But this is exactly the triple whose orientation was used to determine the left-to-right ordering of q relative to r in UNSTRETCH(P). Because p_0 is above the line through q and r in the corresponding stretched configuration, the orientation of (p_0, q, r) will be -1 if q is left of r and $+1$ if q is right of r, matching the orientation of the same triple in S that was used to determine this left-to-right ordering. □

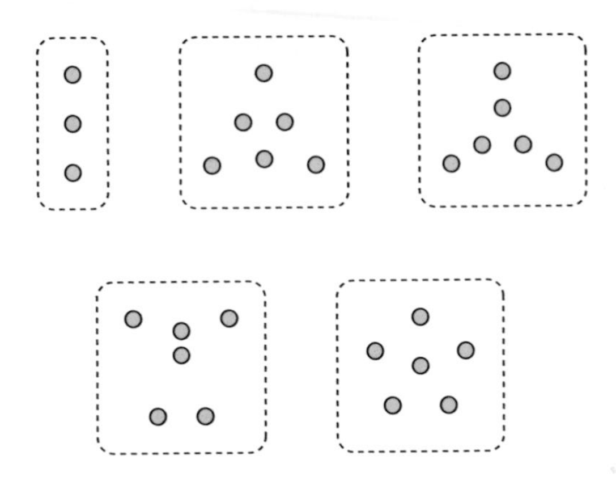

Figure 14.7. Some obstacles for STRETCHED.

Putting Observation 14.17 and Lemma 14.19 together gives us a characterization of the STRETCHED configurations.

Corollary 14.20

The STRETCHED *configurations are exactly the configurations that contain stretched paths.*

Although being a STRETCHED configuration is a monotone property, we have not determined all the obstacles for this property. Every five-point configuration in general position is stretched, but (as well as the three-point line) there exist several six-point obstacles; see Figure 14.7.

Open Problem 14.21

Does STRETCHED have bounded obstacles?

A positive answer to the previous question would also provide a positive answer to the next one, via Theorem 6.8.

Open Problem 14.22

Does STRETCHED have sample-based property testing algorithms with sublinear sample size, constant false positive probability, and constant distance from the property?

14.5 Recognition

Despite not fully understanding the obstacles for STRETCHED, we can recognize stretched configurations efficiently by using stretched paths.

Theorem 14.23

We can test whether configuration S is stretched, and if so find a permutation π such that $S = \text{STRETCH}(\text{PLOT}(\pi))$, in polynomial time.

Proof. We follow the algorithm outlined below.

Algorithm 14.24 (Recognition of stretched configurations)

1. Check that S is in general position, and if not return that it is not stretched.
2. For each point p on the convex hull of S, perform the following steps, until all points have been tested or a stretched path has been found.
   a. Start constructing a sequence P of points, initially consisting only of p.
   b. While there exists at least one point q in S but not in P such that line segment pq is not crossed by any line through two points r and s that are also in S but not in P, add q to the end of P. If there is more than one choice for q, select one arbitrarily.
   c. If P contains all points of S, then it is a stretched path and we break out of the loop over convex hull points p; otherwise, continue on to the next convex hull point.
3. If this search found a stretched path P, then S is stretched, and the desired permutation π can be constructed as UNSTRETCH(P). Otherwise, S is not stretched.

The path P found by this algorithm is stretched by construction, so when the algorithm claims that S is stretched and returns a permutation, it is necessarily correct. However, we must also show that, when there exists a stretched path P', the algorithm will necessarily also find a stretched path P (not always equal to P'). The key properties needed to ensure this are the following.

- Unless the algorithm finds another stretched path first, it will test the starting point of the stretched path P' that we have assumed to exist. This is because it tests all convex hull vertices until finding a stretched path, and the starting point of P' is a convex hull vertex.

- For every two points q and r of S, no line through two points later than q and r in path P' can cross segment qr. This is because if this segment were crossed, then so would be at least one of the segments between consecutive pairs of points in P' between q and r. But this would contradict the assumption that P' is stretched.
- Once a point q becomes available to be added to the sequence P in Step 2.b it remains available until it is actually added.[4] For, suppose that P ends at a point s, that sq is uncrossed (so that q can be added next), but that the algorithm actually chooses a different point t to add next. Because both sq and st are uncrossed by the lines through the remaining points, tq is also uncrossed, because a line cannot cross one side of a triangle without crossing one of the other two sides.
- While the algorithm is testing the starting point p of path P', and at any point within the inner loop of Step 2.b, let q be the first point of P' that is not yet included in P. Then, q is necessarily available to be added to P at the next iteration of the inner loop. To see why this is so, let r be the point that was most recently added to P among the points earlier than q in P'. Then at the earlier iteration of Step 2.b that added r to P, the points that would have remained outside P could only be q itself and a subset of the points that are after q in P'. Therefore, at this earlier iteration, line segment rq must have been uncrossed (the second property above), making q available to be added to P. But none of the later iterations could have caused q to become unavailable (the third property above). So it is still available.

Because the last of these properties proves that there will always be a point available to add to the sequence P until all points are used, the algorithm will successfully terminate whenever there exists a stretched path. □

14.6 Finding Subconfigurations

Theorem 7.3 suggests that testing whether one configuration is a subconfiguration of another is unlikely to be fixed-parameter tractable in the size of the smaller configuration. Theorem 7.7 suggests that, similarly, some parameters with finite obstacle size are unlikely to be fixed-parameter tractable. But the stretched configurations provide a rich class of general-position configurations in which subconfiguration testing and parameter computation is easier.

To prove this, we use the correspondence between subconfigurations of stretched configurations and permutation patterns, and the following result on permutation patterns.

[4] This is the defining property of a mathematical structure known as an *antimatroid*.

Theorem 14.25 (Guillemot and Marx, 2014)

Testing whether one permutation τ is a pattern of another permutation π can be done in an amount of time linear in the length of π and fixed-parameter tractable in the length of τ.

By translating this result into the language of configurations, we obtain the following theorem.

Theorem 14.26

Testing whether a configuration T is a subconfiguration of a STRETCHED *permutation S can be done in an amount of time polynomial in $|S|$ and fixed-parameter tractable in $|T|$.*

Proof. We begin by finding a representation of S as

$$S = \textsc{stretch}\big(\textsc{plot}(\pi)\big)$$

for some permutation π, in polynomial time, by Theorem 14.23. We also find a collection C of all of the permutations τ that can represent T:

$$C = \big\{\tau \mid T = \textsc{stretch}\big(\textsc{plot}(\tau)\big)\big\}.$$

There are at most $|T|!$ such permutations, a large number but one that is still fixed-parameter tractable. Then, we apply Theorem 14.25 to test whether at least one member of C is a pattern of π. By Observation 14.9, this will be true if and only if T is a subconfiguration of S. □

Corollary 14.27

Every monotone parameter with finite obstacle size is (nonuniform) fixed-parameter tractable on the stretched configurations.

15 Configurations from Graphs

Many results on the hardness of subconfiguration-related problems can be based on graph drawing algorithms that create configurations from the vertices and edges of a graph. The value of a monotone parameter of configurations, on a configuration generated from a graph, can often provide useful information about the graph that the configuration came from. If this information would allow us to solve a graph problem that is known to be hard, then computing the parameter must be hard as well.

15.1 Definitions

In this section we briefly summarize the terminology from graph theory that we need here.

Definition 15.1 (graphs, vertices, and edges)

A *graph*, for our purposes, consists of a pair (V, E) where V is a finite set and E is a set of unordered pairs of elements of V. The elements of V are called *vertices* (a single one is a *vertex*) and the elements of E are called *edges*.[1] The two vertices of each edge are called its *endpoints*. A vertex and edge are *incident* if the vertex is an endpoint of the edge. Two vertices are *adjacent* if they are both the endpoints of the same edge.

[1] This definition prevents graphs from having infinitely many vertices or edges, from having more than one edge between the same two vertices, or from having edges attached to only one vertex. Other definitions of graphs allow these things, but we will not need their added generality.

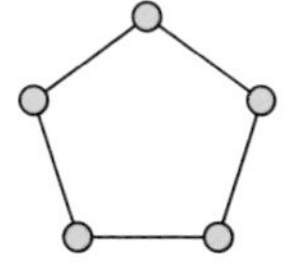 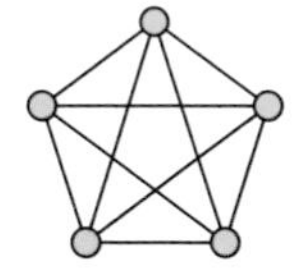 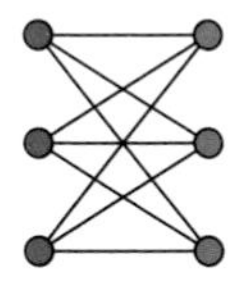

Figure 15.1. Drawings of a cycle graph (left), complete graph (center), and complete bipartite graph (right). The coloring of the complete bipartite graph's vertices shows their partition into two subsets.

Definition 15.2 (graph drawings and planar graphs)

A *drawing* of a graph is a representation of the vertices by points in the plane and the edges by curves from one endpoint to the other that avoid the other vertices. A *crossing* in a drawing is a point where two edges intersect, other than at a shared endpoint of the edges. A drawing is *planar* if it has no crossings, and a graph is a *planar graph* if it can be drawn without crossings. Although, as mathematical objects, the vertices of a drawing are points, we will represent them in figures by small disks, as we do for other kinds of points (Figure 15.1).

Definition 15.3 (paths and connectivity)

A *path* in a graph is an alternating sequence of vertices and edges, starting and ending with vertices, and with each consecutive pair of a vertex and an edge incident to each other. A graph is *connected* if every two vertices can be the first and last vertex of at least one path. An *isolated vertex* is a vertex that has no incident edges.

Definition 15.4 (subgraphs, cliques, and independent sets)

A *subgraph* of a graph is another graph, formed from subsets of the vertices and edges of the first graph. An *induced subgraph* of a given graph and a subset of its vertices is formed by including all the edges that have both endpoints in the vertex subset. A graph or subgraph is *complete* if every pair of vertices in it is adjacent (Figure 15.1, center), and *independent* if no pair of vertices in it is adjacent. A complete subgraph may also be called a *clique*, and a k-clique is a clique with exactly k vertices. A *maximum clique* is a clique with the most vertices possible, and a *maximum independent set* is an independent set with the most vertices possible. A *vertex cover* is

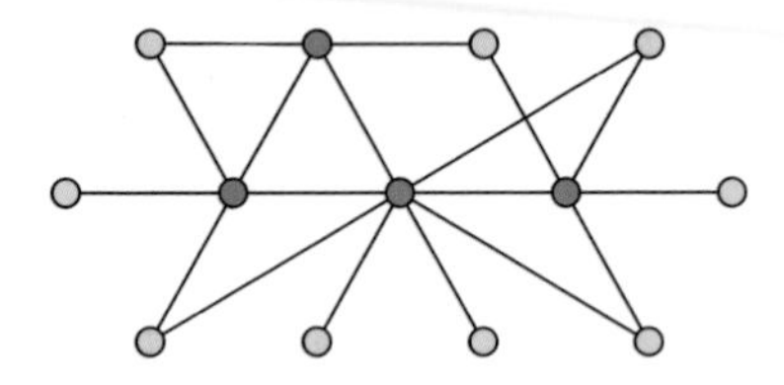

Figure 15.2. Partition of the vertices of a graph into a maximum independent set (yellow) and a minimum vertex cover (blue). No two yellow vertices are adjacent; every edge has a blue endpoint.

a set of vertices that includes at least one endpoint of every edge, and a *minimum vertex cover* is a vertex cover with the fewest vertices possible. The *independence number* is the number of vertices in a maximum independent set, and the *vertex cover number* is the number of vertices in a minimum vertex cover. The vertices of any graph can be partitioned into a maximum independent set and a minimum vertex cover (Figure 15.2).

Definition 15.5 (cycles, regularity, and bipartiteness)

A graph is *bipartite* if its vertices can be partitioned into two subsets U and V such that every edge has one endpoint in each subset. A graph is *regular* if every two vertices have the same number of incident edges, and r-regular if this number is r. A *cycle graph* is a 2-regular connected graph (Figure 15.1, left). Cycle graphs with an even number of vertices are always bipartite. The bipartite graphs are exactly the graphs that do not contain a cycle subgraph with an odd number of vertices. A *complete bipartite graph* is a bipartite graph in which every pair of a vertex from U and a vertex from V is adjacent (Figure 15.1, right). A *Hamiltonian cycle* is a cycle subgraph that includes all the vertices of a given graph.

Definition 15.6 (isomorphism)

An *isomorphism* between two graphs is a bijection of their vertices that preserves adjacency: two vertices are adjacent in the first graph if and only if their images are adjacent in the second graph. Two graphs are *isomorphic* if they have an isomorphism.

Some of our constructions of configurations from graphs will depend not just on the isomorphism type of the given graph, but on an ordering of the vertices of the graph.

Definition 15.7 (ordered graphs)

An *ordered graph* is an undirected graph G whose vertices have been totally ordered. An ordered graph H is an *ordered subgraph* of an ordered graph G if there is an order-preserving bijection between the vertices of H and a subset of the vertices of G such that, for every edge e of H, the image of the endpoints of e under this bijection is a pair of endpoints of an edge in G. More strongly, H is an *ordered induced subgraph* of G if there is an order-preserving and adjacency-preserving bijection between the vertices of H and a subset of the vertices of G (that is, two vertices are adjacent in H if and only if their images are adjacent in G).

15.2 Convex Embeddings of Arbitrary Graphs

The following construction transforms ordered graphs into configurations. For technical reasons, we restrict it to graphs that have no isolated vertices.

Definition 15.8

Let G be an ordered graph with no isolated vertices. Then we define the configuration CONVEX-EMBEDDING(G) to be the result of the following sequence of construction steps.

Algorithm 15.9 (construction of CONVEX-EMBEDDING)

1. For each vertex v of G, place a point for v onto a convex curve C, with the clockwise ordering of the vertices along the curve equal to the ordering of these vertices in the given ordered graph. Curve C should be shallow enough that each three of these points form a right or obtuse triangle. (For instance, a semicircle will work.)
2. Draw a circle of small radius around each vertex. The circle around a vertex v should be small enough that its tangent line L, through the point where the circle is crossed by the line from v to the last vertex, separates v and all earlier vertices from all later vertices. (This is the reason for the right or obtuse triangle requirement in the first step.) Additionally, these circles should be small enough that no single line touches three circles.

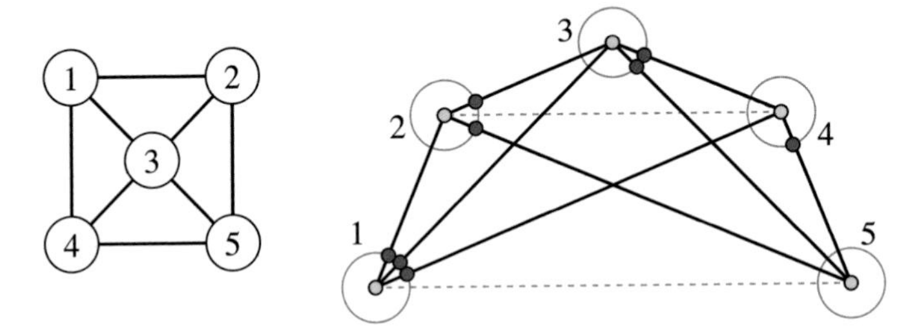

Figure 15.3. A graph G (the five-vertex wheel graph, left) with its vertices ordered according to the numbers shown, and the configuration CONVEX-EMBEDDING(G) (right).

3. For each edge uv of G, where u is earlier and v later in the ordering of vertices of G, place a point at the intersection of the circle for u and line segment uv.

Figure 15.3 depicts an example of the CONVEX-EMBEDDING construction.

The constraints on the angles and circle sizes in this construction ensure that, although there are some arbitrary choices in the exact placement of its points, the order type of the resulting set of points is completely determined.

- Each edge of G corresponds to a three-point line in the resulting configuration, and there are no collinearities other than these three-point lines.
- The points of the configuration that come from vertices can be distinguished from the points that come from edges, because the vertex points are the outer points of their three-point lines and the edge points are the middle points of their three-point lines.

By using rational coordinates for the points on C and rational radii for the circles, we can ensure that all points of CONVEX-EMBEDDING(G) can be realized with rational coordinates or, by scaling the configuration by an appropriate factor to clear denominators, with integer coordinates.

Observation 15.10

If G and H are two ordered graphs, neither of which have isolated vertices, then H is an ordered subgraph of G if and only if CONVEX-EMBEDDING(H) *is a subconfiguration of* CONVEX-EMBEDDING(G).

Proof. Removing vertices or edges from G does not change the construction of CONVEX-EMBEDDING for the remaining vertices or edges, so every ordered subgraph of G corresponds to a subconfiguration in this way. Conversely, a subconfiguration S of CONVEX-EMBEDDING(G) can itself be formed by the CONVEX-EMBEDDING construction only if every point of S belongs to a three-point line in S, in which case we can identify the vertices and edges of a subgraph of G, as above, from the positions of these points within their three-point lines. □

15.3 Hardness Proofs Based on Convex Embeddings

Proof of Theorem 7.3. Recall that the theorem states the NP-completeness and W[1]-hardness of testing whether one configuration of size k (for variable k) is a subconfiguration of another of size n, and also states that it cannot be solved in time $n^{o(\sqrt{k})}$ unless the exponential time hypothesis is false.

We reduce from the problem of finding a complete graph $H = K_c$ as a subgraph of a larger graph G. This problem is known to be NP-complete and W[1]-hard. Additionally, unless the exponential time hypothesis is false, this problem cannot be solved in time $O(|V(G)|^{o(c)})$.[2]

For this problem, the vertex orderings on the two graphs are irrelevant: H is a subgraph of G if and only if, for any arbitrarily chosen pair of vertex orderings, CONVEX-EMBEDDING(H) is a subconfiguration of CONVEX-EMBEDDING(G). We let

$$n = \left|\text{CONVEX-EMBEDDING}(G)\right| = O\left(|V(G)|^2\right)$$

and

$$k = \left|\text{CONVEX-EMBEDDING}(K_c)\right| = O\left(c^2\right).$$

The result follows by a change of variables from the graph numbers of vertices to these configuration sizes. □

Proof of Theorem 7.4. Recall that Theorem 7.4 states that testing whether AVOIDS(C; S) $\geq k$, when both C and S are variable, is Σ_2^{P}-complete. To prove this, we describe a reduction from a known Σ_2^{P}-complete problem, testing whether a given undirected graph G can be made to avoid having a c-vertex complete subgraph by the removal of k vertices (for variable c and k). This problem, a special case of a problem called "generalized node deletion," is listed as Σ_2^{P}-complete by Schaefer and Umans (2002, problem number GT10). The reduction is to arbitrarily order the vertices of G and, letting K_c denote the c-vertex complete graph, construct

$$C = \text{CONVEX-EMBEDDING}(K_c) \text{ and}$$

$$S = \text{CONVEX-EMBEDDING}(G).$$

The ordering on the vertices of K_c does not affect its embedding, nor does the ordering on G affect which ordered cliques are ordered subgraphs of G. Thus, each subconfiguration of C in S corresponds to a complete c-vertex subgraph of

[2] Chen et al. (2006).

G. A subconfiguration of S that avoids C can be formed by removing a collection of vertex and edge points from S, so that the removed points include at least one point from each copy of C. However, if an edge point is removed, we can also avoid C by removing either of the endpoints of the same edge. Therefore, the number of vertices to remove to avoid C can always be minimized by removing only vertex points.

In summary, removing k vertices from G can eliminate all c-vertex complete subgraphs, if and only if removing k points from S can eliminate all copies of C as a subconfiguration of the remaining points. That is, the answer to the given instance of generalized node deletion is yes if and only if

$$\textsc{avoids}(C;\ S) \geq |S| - k.$$

This reduction proves the Σ_2^{P}-hardnes of testing $\textsc{avoids}(C;\ S) \geq k$. The problem is in Σ_2^{P}, as one can solve it by a brute force search of all k-point subconfigurations, testing for each one whether it avoids C. Therefore, it is Σ_2^{P}-complete. □

Proof of Theorem 7.7. Recall that Theorem 7.7 states the existence of a W[1]-hard parameter with polynomial obstacle size. This is true of the size of the largest subconfiguration of the form $\textsc{convex-embedding}(K_c)$. The obstacles for this parameter are the configurations $\textsc{convex-embedding}(K_c)$, whose size is equal to the parameter value being tested. □

Proof of Theorem 9.3. Recall that Theorem 9.3 states that it is NP-hard and APX-hard to compute or approximate MAX-GENERAL and DELETE-TO-GENERAL.

We prove this by a reduction from maximum independent set in graphs. For any graph G, order the vertices of G arbitrarily, and construct the configuration $S = \textsc{convex-embedding}(G)$. Then $\textsc{max-general}(S) = m + \alpha(G)$, where m is the number of edges and α the independence number of G, and $\textsc{delete-to-general}(S)$ equals the vertex cover number of G.

Both the independence number and the vertex cover number are NP-hard for arbitrary graphs, and APX-hard for sparse graphs (graphs constrained to have a number of edges at most some fixed constant times the number of vertices). Because these graphs always have independent sets of linear size, computing $m + \alpha(G)$ is also APX-hard. Therefore, applying this reduction to sparse graphs shows that both MAX-GENERAL and DELETE-TO-GENERAL are NP-hard and APX-hard, as claimed. □

15.4 Three-Line Embeddings of Bipartite Graphs

The next proofs use a similar construction of a configuration from a graph. However, by using a more restrictive class of graphs it obtains a more structured class of configurations.

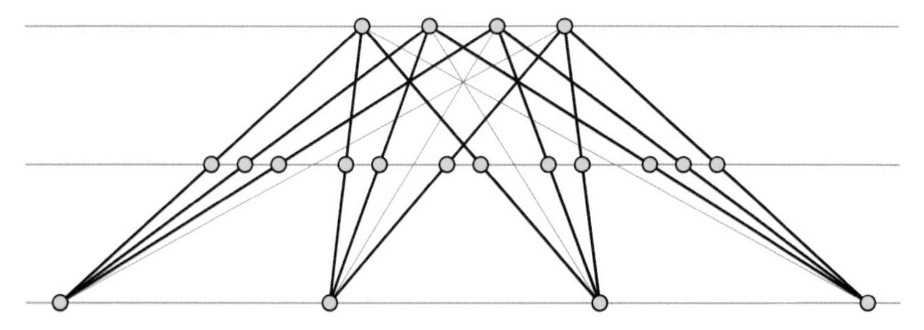

Figure 15.4. Three-line drawing THREE-LINE(Q), where Q is the graph of a cube.

Definition 15.11

We say that a bipartite graph $G = (U, V, E)$ is ordered when each of its two vertex sets U and V has an associated linear ordering. We construct a configuration THREE-LINE(G) from any order bipartite graph $G = (U, V, E)$, by proceeding as follows.

- Draw three horizontal lines equally far apart (blue in Figure 15.4).
- On the upper line, place a point for each vertex of U, with the left-to-right ordering of these points matching the linear ordering of U. These are the upper four yellow points of Figure 15.4, four of the eight vertices of a cube. The spacing of these points can be arbitrary.
- Let ℓ be the farthest distance between any two of the points on the upper line.
- On the lowest line, place a point for each vertex of V, with the left-to-right ordering of these points matching the linear ordering of U. Space each two consecutive points farther than ℓ apart. These are the lower four yellow points of Figure 15.4, the other four vertices of the cube.
- For each edge $u_a v_b$ (where $u_a \in U$ and $v_b \in V$) place a point where segment $u_a v_b$ crosses the middle horizontal line. The figure shows the edges of the cube as black and the points placed in this step as the 12 yellow midpoints of the black edges.

Then THREE-LINE(G) is the configuration of the point set constructed in this way.

In this construction, the wider spacing on the bottom line implies that the line segments connecting pairs of points from the two lines (red and black in the figure) cross the middle horizontal line in a consistent order. Segment $u_a v_b$ will cross to the left of $u_c v_d$ (where u_a, $u_c \in U$ and v_b, $v_d \in V$) if v_b is earlier than v_d, or if $v_b = v_d$ and u_a is earlier than u_c.

As with CONVEX-EMBEDDING(G), with some technical restrictions we can reconstruct the graph G from the order-type of THREE-LINE(G).

Lemma 15.12

Let $G = (U, V, E)$ be an ordered bipartite graph with no isolated vertices. Suppose also that G contains a path v_1–u–v_2 where $u \in U$ and where v_1 and v_2 are nonconsecutive in the ordering of V. Then G, its bipartition (U, V), and the orderings on U and V can be uniquely recovered from THREE-LINE(G). *No other graph meeting the same conditions can give rise to the same three-line configuration.*

Proof. Note that G must have at least one edge other than the ones in the path v_1–u–v_2, incident to the vertex between v_1 and v_2. Therefore, the middle of the three initial lines used to construct THREE-LINE(G) is heavy. It is the only heavy line that contains the interior points of two other nondisjoint heavy lines (the lines for edges v_1–u and u–v_2). The edge points of the configuration are the ones on the middle initial line and the vertex points are the ones not on it.

At this point, we can recover G as an unordered graph: it is a graph with a vertex for each vertex point of the configuration, in which two vertices are adjacent if they form a three-point line with one of the edge points. Next, we must determine which side of the edge line is U and which is V. To do so, we use the path v_1–u–v_2 again. Because of this path, at least one vertex u in U belongs to three-point lines with a set of edge points that are nonconsecutive along the edge line. However, all of the vertices in V belong to three-point lines with sets of edge points that are consecutive. This difference allows us to determine which side of the edge line is U, and which is V.

The ordering of the vertices in U and V can be recovered from the ordering of the corresponding points along their lines, using Lemma 3.10. □

Corollary 15.13

If two ordered bipartite graphs G and H both meet the conditions of Lemma 15.12, then THREE-LINE(H) *is a subconfiguration of* THREE-LINE(G) *if and only if H is a subgraph of G with the same vertex ordering as G.*

We will use this construction to prove non–well-quasi-ordering of classes of configurations that include all three-line graph embeddings, and to prove lower bounds on the numbers of configurations in these classes.

Lemma 15.14

The configurations of the form THREE-LINE(G) *are not well-quasi-ordered, and the number of distinct configurations of this type with n vertices is exponential in* $n \log n$.

Proof. If F is an antichain in the partial order of bipartite graphs and their subgraphs, and we order the vertices of each graph of F arbitrarily, meeting the conditions of Lemma 15.12, then by Corollary 15.13 the three-line configurations of the resulting ordered graphs will form an antichain in the partial order of configurations and subconfigurations. So choose F to be the family of all cycle graphs whose length is even and at least six. We observe that, for any ordering of such a graph, there will be at least one vertex in U whose two neighbors are nonconsecutive in V, for the $2n$-vertex cycle will have n vertices in U and only $n-1$ consecutive pairs in V. Therefore, these graphs give an infinite antichain in the configurations THREE-LINE(G).

To prove that the number of configurations of this type is exponential in $n \log n$, we begin by making the simplifying assumption that n is a multiple of four, so that we can form an n-point configuration as THREE-LINE(G) where G is an $n/2$-vertex cycle graph. For such a cycle C, number the vertices of C from 1 to $n/2$; then we can represent an ordering of G as a permutation of the odd-numbered vertices (as the set U) together with a separate permutation of the even-numbered vertices (as the set V). This representation is not unique, however, because there are $n/2$ choices for which vertex to number as 1 and two choices for which way around the cycle to extend the numbering. Therefore, the number of nonisomorphic orderings of the $n/2$-vertex cycle graph is exactly

$$\frac{(n/4)!^2}{n}.$$

The same formula, which is exponential in $n \log n$, counts the number of configurations THREE-LINE(G) where G is an ordered $n/2$-vertex cycle.

If n is not a multiple of four, we use the same construction for graphs consisting of an even cycle of length at least six plus up to three isolated edges. Each isolated edge adds three points to THREE-LINE(G), allowing the remaining number of points corresponding to the cycle part of the graph to be a multiple of four. □

Proof of Theorem 8.16, part II. The second part of this theorem states that the configurations that can be covered by three parallel lines are not well-quasi-ordered and that the number of such configurations with n points is exponential in $n \log n$. This follows immediately from Lemma 15.14 and from the fact that the configurations THREE-LINE(G) can (by construction) be covered by three parallel lines. □

Proof of Theorem 9.7 and Theorem 11.22, part II. The second parts of Theorem 9.7 and Theorem 11.22 concern the non–well-quasi-ordering of the configurations for which MAX-GENERAL(S) < 7 and for which MAX-CONVEX(S) < 7, respectively. They also state that the number of these configurations that have n points is exponential in $n \log n$. These claims, for both theorems, follow immediately from Lemma 15.14 and from the fact that (in either case) these configurations include all configurations of the form THREE-LINE(G). □

The same construction can also be used to prove analogues of Theorem 7.3, according to which testing whether one configuration is a subconfiguration of another is NP-hard and not fixed-parameter tractable (unless the exponential time hypothesis is false), even when both configurations have LINE-COVER ≤ 3, MAX-GENERAL < 7, or MAX-CONVEX < 7. The proof of Theorem 7.3 involved searching for subconfigurations of the form CONVEX-EMBEDDING(K_c), where K_c is a complete graph. Instead, to show hardness for these restricted classes of configurations, we search for subconfigurations of the form THREE-LINE($K_{c,c}$), where $K_{c,c}$ is a complete bipartite graph. We omit the details.

16 Universality

A *planar graph* is a graph that can be drawn in the plane with its vertices as points and its edges as noncrossing curves (often drawn as line segments). Universal point sets are point sets that can form the vertex set of any sufficiently small planar graph. Before defining this concept more carefully as the monotone parameter UNIVERSAL (Definition 16.1), we look at a computer puzzle that has popularized some of these concepts.

16.1 Planarity

Planarity[1] is a computer puzzle game based on planar graph drawing, originally devised by John Tantalo and Mary Radcliffe. In it, one is presented with a tangled drawing of a graph (Figure 16.1). The goal is to untangle it, by moving the vertices one by one, until the result has no crossings. As the vertices move, the edges move with them as straight line segments. Each level of the puzzle presents a more complicated graph to be untangled. Although there exist computer algorithms that can solve these puzzles efficiently, their visual complexity nonetheless makes them a challenge to human puzzle-solvers.

The original version of this puzzle generates its graphs from arrangements of randomly generated lines (Figure 16.2, left). However, the sharp angles of these arrangements make it difficult to reconstruct them while solving the puzzle. Indeed, even finding an arrangement that generates a given graph is a hard problem, much harder than finding an untangled drawing for the graph.[2]

[1] Online at http://planarity.net. [2] As Shor (1991) shows, it is $\exists\mathbb{R}$-complete.

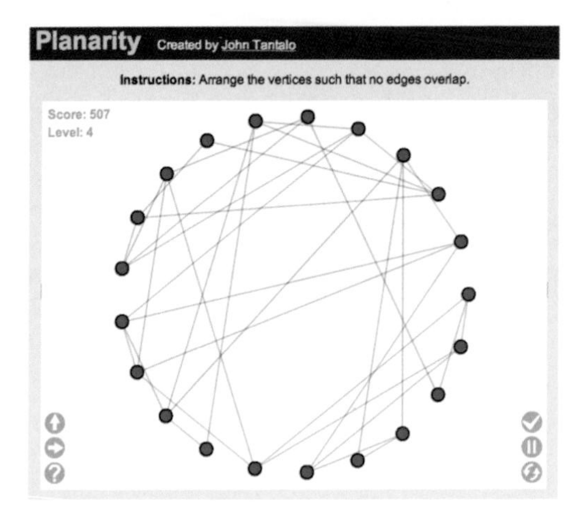

Figure 16.1. Initial state of Planarity, from Eppstein (2014).

Therefore, it can be easier to search for a solution to the puzzle that places the vertices on a grid (or a rough human approximation to a grid), such as the one in Figure 16.2, right. But when we're just starting the puzzle, we don't know what the whole grid drawing will look like, if it even exists. How big should we make the grid squares? If they're too big, the drawing won't fit in the window, and we'll have to waste time and moves shrinking it. But if they're too small, it will be more difficult to precisely place each of the graph vertices into its grid position.

Eppstein (2014) gives an answer to this grid sizing question. He shows that the line arrangement graphs used by Planarity can always be drawn without crossings on a grid of dimensions $O(n^{1/2}) \times O(n^{2/3})$, where n is the number of vertices in the graph. More, there is a subconfiguration of only $O(n \log n)$ points from this grid that can still support a drawing of one of these arrangement graphs. This subconfiguration has a simple structure: if the grid has $O(n^{1/2})$

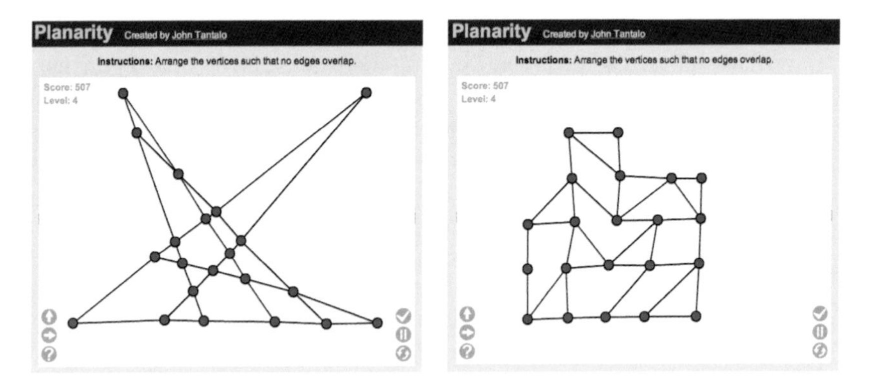

Figure 16.2. Two solutions to the puzzle of Figure 16.1, from Eppstein (2014).

rows and $O(n^{2/3})$ columns, its subconfiguration is formed from taking a smaller number of points from each row (it doesn't matter which ones), where the numbers of points to take from each row are computed from the sequence

$$1, 3, 1, 7, 1, 3, 1, 15, 1, 3, 1, 7, 1, 3, 1, \ldots$$

by multiplying each sequence value by $O(\sqrt{n})$.

However, many clones of the Planarity game are available online and for download, not all of which generate their planar graphs in the same way. The bounds above apply only to the original Planarity game and its line-arrangement graphs. For planar graphs generated in other ways, can we still always draw them on grids? How big a grid is needed? Can we find other nongrid configurations that allow us to draw any n-vertex graph with fewer points? Although we know only partial answers to these questions, we can formalize them using the theory of universal point sets.

16.2 Definition

We now describe more carefully what it means to draw a graph using a fixed set of vertex positions, and use this concept to define a monotone parameter that captures the concept of universal point sets.

Definition 16.1

A *planar straight-line drawing* of a graph G is mapping from the vertices of G to distinct points in the plane such that the closed line segments connecting adjacent pairs of vertices do not have any intersection points other than shared endpoints. This implies that edges cannot cross or overlap on the same line and that an endpoint of one edge cannot lie on another edge. A configuration S *supports* a planar graph G if G has a straight-line drawing all of whose vertices belong to S.

Complicated graphs can sometimes be supported by unexpectedly simple configurations, as the following example shows.

Example 16.2

The $n \times n$ grid graph is the graph formed by placing a vertex at each point (i, j) with $1 \leq i, j \leq n$, and by connecting two such vertices by an edge when they are at distance one from each other. Thus, it can naturally be supported by the configuration GRID(n, n). However, although this representation has

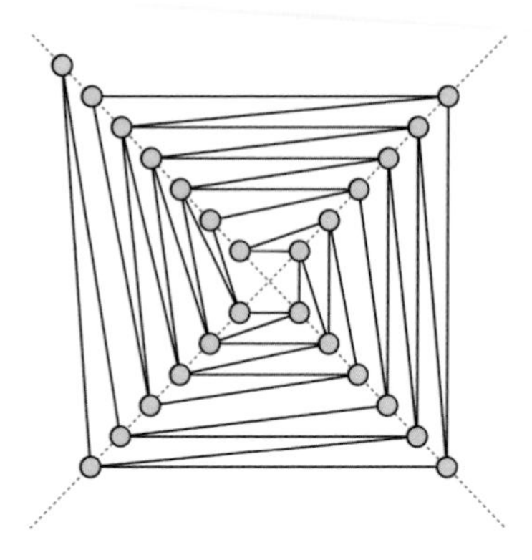

Figure 16.3. A planar straight-line drawing of a 5×5 grid graph, supported by a configuration S with LINE-COVER$(S) = 2$.

LINE-COVER$(S) = n$, the same graph can also be supported by configurations S with LINE-COVER$(S) = 2$ (Figure 16.3).[3]

Universal configurations must support all sufficiently small graphs, not just a single graph.

Definition 16.3

A configuration S is *universal* for the n-vertex planar graphs if it supports every planar graph with at most n vertices. We define the monotone parameter UNIVERSAL(S) to be the maximum value of n such that S is universal for the n-vertex planar graphs.

Example 16.4

For collinear configurations S with more than one point, UNIVERSAL$(S) = 2$, because the triangle graph cannot be drawn on these configurations. For configurations of three or more points in convex position, UNIVERSAL$(S) = 3$, because these point sets cannot be used to draw the planar graph K_4, the complete graph on four vertices.

The configuration TETRAD of Figure 3.3 is universal for four-vertex planar graphs, because every such graph is a subgraph of the four-vertex complete

[3] For a closely related style of layout on three rays instead of two crossing lines, see Bannister et al. (2016).

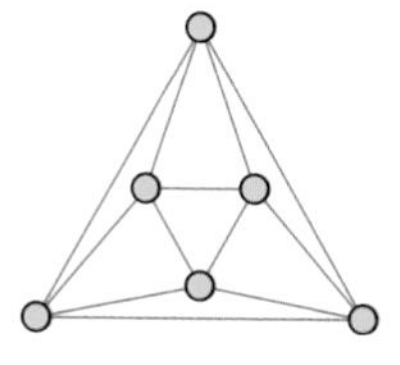
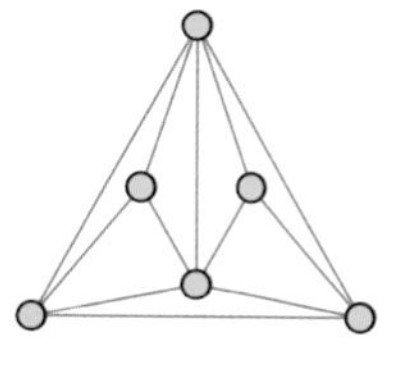

Figure 16.4. SAWTOOTH(3) can be used to draw both of the possible maximal planar graphs on six vertices (Example 16.5).

planar graph K_4, and K_4 can be drawn without crossings on the tetrad. Thus, UNIVERSAL(TETRAD) $= 4$.

16.3 Obstacles

The main unsolved question about universal configurations asks how small they can be. Despite the efforts of multiple researchers, the known upper and lower bounds on the size of universal configurations are far apart: only linear lower bounds are known,[4] but the smallest known universal configurations have quadratic size.[5]

If a configuration S is universal for the n-vertex planar graphs, and cannot be reduced to a smaller subconfiguration while preserving its universality, then it is an obstacle for UNIVERSAL. Therefore, we can approach the question of how small a universal configuration can be by studying this parameter's obstacles.

Example 16.5

There are two distinct *maximal* six-vertex planar graphs. These are the graphs to which no more edges can be added while preserving the planarity of the graph. Both of these graphs can be drawn using SAWTOOTH(3) as the vertex set, as shown in Figure 16.4. Every other planar graph with at most six vertices can be drawn by removing edges or vertices from one of these two drawings. Thus,

$$\text{UNIVERSAL}\big(\text{SAWTOOTH}(3)\big) = 6.$$

However, removing any point from SAWTOOTH(3) leaves a five-point configuration, not enough to draw a six-vertex graph. Thus, SAWTOOTH(3) is minimal for its value of UNIVERSAL and is an obstacle for UNIVERSAL. It forms a universal configuration for drawing the 6-vertex planar graphs, and is one of the smallest possible universal configurations for these graphs.

[4] de Fraysseix et al. (1988); Chrobak and Karloff (1989); Kurowski (2004).
[5] de Fraysseix et al. (1988); Schnyder (1990); Brandenburg (2008); Bannister et al. (2014).

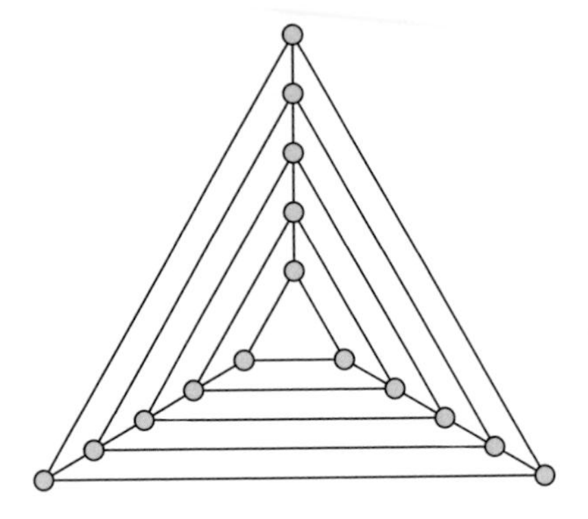

Figure 16.5. The nested triangles graph.

For all $n \leq 10$, there exist universal configurations for n-vertex planar graphs, with exactly n points. For $n \geq 15$, the property of being universal for n-vertex graphs requires more than n points, but it remains unknown how many points are needed as a function of n.[6] The question of the size of universal configurations can be formulated as the following problem.

Open Problem 16.6

How big is the smallest obstacle for the property that UNIVERSAL$(S) < k$?

One way to construct universal configurations is to use grids. This works because every planar graph can be drawn using large-enough integer coordinates. In other words, large-enough grid configurations are universal. For instance, both GRID$(n-1, n-1)$ and

$$\text{GRID}\left(\left\lceil \frac{4n}{3} \right\rceil, \left\lceil \frac{2n}{3} \right\rceil\right)$$

are universal for n-vertex planar graphs. For both of these grids, one can reduce the size of the configurations by another factor of two by using only a triangular subconfiguration of the grid.[7] However, every universal grid must have both of its dimensions proportional to the number of vertices,[8] because otherwise it would not support the *nested triangles graph* shown in Figure 16.5.

The construction that generates the smallest known universal configurations is based not on grids, but on STRETCHED permutations (Chapter 14).

[6] Cardinal et al. (2015).
[7] de Fraysseix et al. (1988); Schnyder (1990); Brandenburg (2008).
[8] Dolev et al. (1984).

Theorem 16.7 (Bannister et al., 2014)

The n-vertex planar graphs have universal configurations of size

$$\frac{n^2}{4} - \Theta(n).$$

In outline, the proof of this result takes the following steps.

- The authors prove that every n-vertex planar graph G has a 213-avoiding permutation π_G on the numbers from 1 to n, such that $\textsc{stretch}\big(\textsc{plot}(\pi_G)\big)$ supports G.
- They find a *superpattern* of the length-n 213-avoiding permutations, of length $n^2/4 + O(n)$. This is another permutation σ that contains all of the length-n 213-avoiding permutations as patterns.
- Therefore,

  $$\textsc{universal}\left(\textsc{stretch}\big(\textsc{plot}(\sigma)\big)\right) \geq n.$$

  This universal configuration has slightly more points than the bound in the theorem, $+\Theta(n)$ rather than $-\Theta(n)$. But by observing that the permutations π_G all have three values in common, and using superpatterns for length-$(n-3)$ permutations plus three additional points for these three values, the authors achieve the stated bound.

Unfortunately, this only improves on the size of the grid constructions for universal configurations by a constant factor. Additionally, because of the stretching transformation that it uses, the graph drawings constructed in this way are not likely to be very readable.

Although not as central to this subject, it is also of interest to know how large the obstacles for $\textsc{universal}$ can be. For instance, if the largest obstacles are small, this could help compute the value of $\textsc{universal}(S)$ for an input configuration S more quickly.

Observation 16.8

Parameter $\textsc{universal}(S)$ *has finite obstacle size.*

Proof. Suppose S is an obstacle for $\textsc{universal}(S) < k$. That is, S is a minimal universal configuration for the k-vertex planar graphs. There are exponentially many k-vertex planar graphs.[9] For each one, find a drawing of it using vertex

[9] Giménez and Noy (2009).

positions in S. Then the whole obstacle S must be the union of the selected positions, for otherwise we could remove any unused points, contradicting the assumed minimality of the obstacle. Therefore, S has at most exponentially many points. □

This raises the following question.

Open Problem 16.9

Does UNIVERSAL have polynomial obstacle size?

Additionally, we know very little about the computational complexity of this parameter. We could test whether $\text{UNIVERSAL}(S) < k$ by generating all k-vertex planar graphs,[10] and for each graph performing a brute force search to test whether it can be drawn using vertices from S. However, although this algorithm is polynomial for fixed k, it is not fixed-parameter tractable.

A hint that the problem of computing UNIVERSAL might be hard is given by the following result.

Theorem 16.10 (Cabello, 2006)

It is NP*-complete to determine, for a configuration S and graph G, whether S supports G.*

The problem remains NP-complete when $|C| = |V(G)|$ and G is 2-vertex-connected. It follows that, for a given S and k, testing whether $\text{UNIVERSAL}(S) < k$ is in the complexity class Σ_2^{P}. This is because it can be expressed as the search for a k-vertex planar graph that S does not support. Testing whether a graph is supported is in NP, so searching for an unsupported graph is in Σ_2^{P}. Jean Cardinal[11] has suggested that testing whether $\text{UNIVERSAL}(S) < k$ is a natural candidate to be Σ_2^{P}-complete.

Open Problem 16.11 (Cardinal)

What is the computational complexity of computing or approximating $\text{UNIVERSAL}(S)$? Is it Σ_2^{P}-complete to test whether $\text{UNIVERSAL}(S) < k$?

[10] Brinkmann and McKay (2007).
[11] Personal communication, 2017, based on an email from Cardinal to Cabello, 2013.

16.4 Relations to Other Parameters

One way to attack Open Problem 16.6 on the size of universal sets would be to show that they have a complicated internal structure, and need many points to support this structure. Toward this end, we show that some other parameters must be large in any universal configuration.

Observation 16.12

UNIVERSAL $\ll$ ONION.

Proof. The nested triangles graph, a graph formed by connecting a sequence of linearly many triangles by adding edges between corresponding vertices of consecutive triangles, requires that (for any drawing) at least half of the triangles in the sequence form nested onion layers. However, it can be drawn (as shown in Figure 16.5) with its vertices on three rays meeting at a point; this configuration has an unbounded number of onion layers but (by Theorem 16.13, below) bounded universality. □

Despite the examples of a grid graph being drawable using points on only two lines (Example 16.2), and the nested triangles graph on three lines (Figure 16.5), some other planar graphs need unbounded numbers of lines, as the next result shows.

Theorem 16.13

UNIVERSAL $\ll$ MAX-GENERAL.

Proof. The configuration POLYGON(n) has bounded values of UNIVERSAL:

$$\text{UNIVERSAL}\big(\text{POLYGON}(n)\big) = 3,$$

because the only planar graphs that can be drawn on it are the *outerplanar graphs* (graphs in which all vertices belong to the unbounded face of a drawing), and the complete graph K_4 is not outerplanar. However, this configuration has unbounded values of MAX-GENERAL:

$$\text{MAX-GENERAL}\big(\text{POLYGON}(n)\big) = n.$$

To complete the proof, we need to show that configurations with small values of MAX-GENERAL (or equivalently and more conveniently LINE-COVER) cannot be universal. A planar graph, variously called an *Apollonian network*, *Kleetope*, or *planar 3-tree*, can be formed by starting with a drawing of a triangle and then performing ℓ levels of subdivision in which we replace each bounded triangle

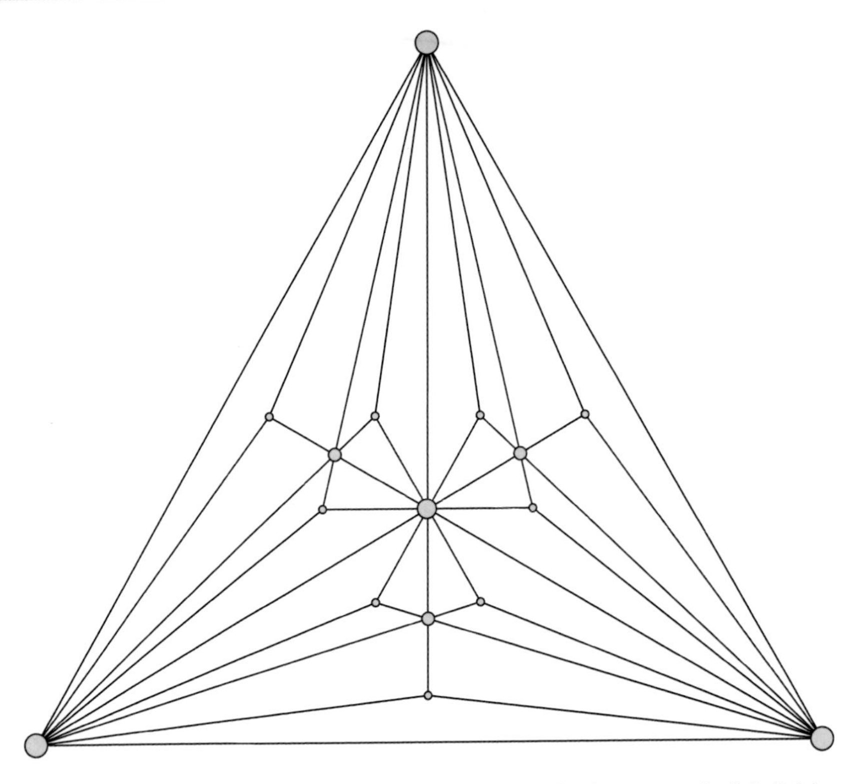

Figure 16.6. Three levels of subdivision of an Apollonian network. Subdivision level is shown by point size. Despite the many collinear triples in this drawing, these graphs require points on an unbounded number of lines.

of the drawing by three triangles meeting at an interior point (Figure 16.6). For any drawing of this graph, and any set of $\ell - 1$ lines, we can find a vertex that is not on any of the given lines. To do so, start at the initial triangle (regardless of whether it is actually the outer face of the given drawing) and then use the following rule to find a sequence of subdivided triangles. The rule is, at any triangle T in the sequence, let L be one of the given lines containing the subdivision point of T, and choose the next triangle in the sequence to be the one of the three subdivisions of T whose interior is disjoint from L. Each subdivision step eliminates one of the lines from further consideration, so by the last step of the sequence we must have found a subdivision point that does not belong to any of the lines.

We can calculate the number of vertices in the ℓ-level Apollonian network as $n = 3 + \sum_{i=0}^{\ell-1} 3^i$. For this n, a point set with fewer than ℓ lines cannot be universal for the n-vertex planar graphs. That is,

$$\text{UNIVERSAL}(S) < 3 + \sum_{i=0}^{\text{LINE-COVER}(S)-2} 3^i,$$

showing that UNIVERSAL is bounded by a function of LINE-COVER. The result follows from the earlier equivalence between LINE-COVER and MAX-GENERAL. □

It is plausible that a similar proof will show that universal point sets for large graphs cannot be covered by few convex polygons, or by few weakly convex polygons, but we do not have a proof.

Open Problem 16.14

What relation, if any, exists between UNIVERSAL and WEAK-PARTITION, or between UNIVERSAL and CONVEX-PARTITION?

17 Stabbing

Many of our positive results on well-quasi-ordering of classes of configurations rely on decompositions into line segments that are not *stabbed* by the lines through pairs of points.

Definition 17.1

Line segment s is stabbed by line ℓ if s and ℓ intersect in a single point, interior to s.

As a warm-up for the concept of stabbing, consider the following puzzle. From the 16 points of GRID(4, 4), form eight disjoint line segments, with the points as their endpoints, that form an *opaque forest*, blocking all lines of sight from one side of the grid to the other.[1] Every line that passes through the interior of the grid without touching any grid points must stab at least one of these segments, for otherwise one could see through the grid along that line. Equivalently, for every nontrivial line partition of the points (Definition 3.11), at least one of the line segments should have an endpoint on each side of the partition. A solution is depicted in Figure 17.10 at the end of the chapter.

[1] The famous opaque forest problem of Mazurkiewicz (1916) asks for curves blocking all lines of sight through a square or other shape and minimizing total length, but here we ask for a different property, that the curves form nonintersecting line segments ending at each grid point.

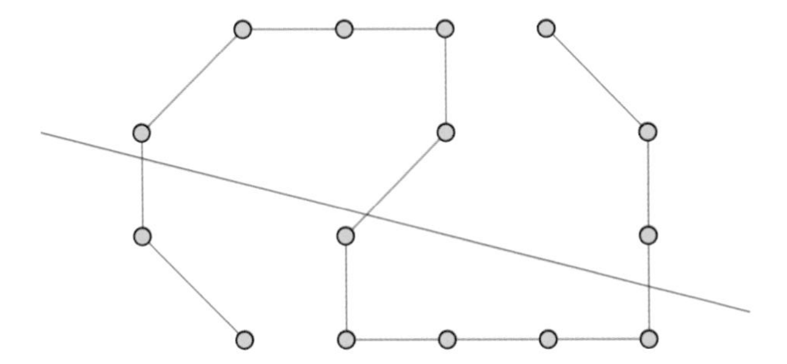

Figure 17.1. A configuration S with PATH-STAB$(S) = 3$, a path through S with stabbing number three (blue), and a line that stabs the path three times (red).

We have seen stabbing already, in the definition of the stretched paths used in Section 14.4 to characterize the STRETCHED configurations. The same idea is also closely related to some early investigations in geometric range searching data structures.

17.1 Path and Tree Stabbing Numbers

Definition 17.2

The *stabbing number* of a collection of line segments is the largest number of segments that can be simultaneously stabbed by a single line. The stabbing number of a graph (with points from a given configuration as its vertices) is the same number applied to the edges of the graph, represented as line segments.

The *path stabbing number* PATH-STAB of a configuration S is the smallest stabbing number of any path through all points of S. The *tree stabbing number* TREE-STAB of a configuration S is the smallest stabbing number of any tree that spans all points of S.

Example 17.3

A configuration S has PATH-STAB$(S) = 1$ when it is collinear, with the path traversing the points of S in the order that they appear on their common line. A configuration S has PATH-STAB$(S) = 2$ when it is in weakly convex position, with the path traversing the points of S in clockwise order around their convex hull. Figure 17.1 shows an example of a configuration S with PATH-STAB$(S) = 3$.

For any line L that stabs the largest number of segments of some given collection C of line segments, perturbing L by a small amount does not change the segments in C that it stabs. Therefore, we may assume that, for any configuration S and any path or tree that spans S, the line that stabs the most segments of the path or tree is disjoint from S. It may not be completely obvious that the path stabbing number and tree stabbing number are invariants of configurations – that is, that perturbing a point set while preserving its order type preserves its stabbing number. However, this invariance follows from Corollary 3.13. This corollary shows that, whenever a line L stabs a set of line segments in a realization of a configuration S, the same set of line segments can also be stabbed by a line in any other realization of S.

Observation 17.4

PATH-STAB *is monotone.*

Proof. Let S be an arbitrary configuration, let P be any path through S that has stabbing number PATH-STAB(S), and let p be an arbitrary point to be removed from S. Then in the resulting subconfiguration, we choose a spanning path P' that passes through the remaining points in the same order. The only lines whose stabbing numbers could increase after this change are the ones that cross path P without stabbing it by passing through point p itself, because in P' such a crossing could be converted into a stabbing. However, for any such line L, perturbing L to avoid p also produces a line of greater stabbing number in the original configuration S and path P. These lines are not the ones that achieve the maximum stabbing number for P. Therefore, the stabbing number for P' is no more than for P, and the subconfiguration of S formed by removing p has path stabbing number no more than for S. □

Open Problem 17.5

Is TREE-STAB monotone?

17.2 Stabbing versus Shattering and Length

Range searching is the problem of setting up data structures that can find the points within a query shape (such as a rectangle) and return some aggregate information about these points (such as how many there are), without having to examine each point of a data set individually. Paths of low stabbing number can be used for several types of range searching problem, although this technique has largely been supplanted by other methods. Several early data structures for range searching were based on the fact that any point set of size n can be spanned by a path or tree of stabbing number $O(\sqrt{n})$.[2]

Algorithm 17.6 (iterated reweighting for low-stabbing-number trees)

A tree with stabbing number $O(\sqrt{n})$ may be obtained by an iterated reweighting technique that performs the following steps.

1. Construct the set of all line partitions of the given configuration S and find a representative line for each partition.
2. Assign weights to these representative lines, initially all equal.
3. While the configuration still has more than one point remaining, repeat the following steps.
   a. Find a line segment s connecting two points of S such that the lines crossed by s have as low total weight as possible.
   b. Remove one endpoint of s from S.
   c. Double the weight of each representative line that crosses s.
4. Return the collection of all segments s found within the loop.

A path with at most double the spanning number of the same tree can be found by making two copies of each edge of the tree to produce an Eulerian multigraph, and then connecting the points into a path in the order of their first appearance within an Euler tour of this multigraph. A version of this algorithm that chooses the factor by which weights are multiplied more carefully (instead of doubling the weights) can find a tree with stabbing number $4\sqrt{n} + o(\sqrt{n})$ and a path with stabbing number $8\sqrt{n} + o(\sqrt{n})$.[3]

To show that this $O(\sqrt{n})$ bound is asymptotically tight, Welzl (1992) constructs point sets with small *shattering number*, defined as follows.

Definition 17.7

Define the *shattering number* $\text{SHATTER}(S)$ of a configuration S to be the minimum number of lines in an arrangement A of lines that are all disjoint from S, such that every two points of S are separated by at least one line of A. Equivalently, the lines of A must stab every line segment determined by two points of S. The fact that this is a monotone parameter follows from Lemma 3.10.

[2] Chazelle and Welzl (1989); Matoušek (1991b); Agarwal (1992); Welzl (1992).
[3] Welzl (1992).

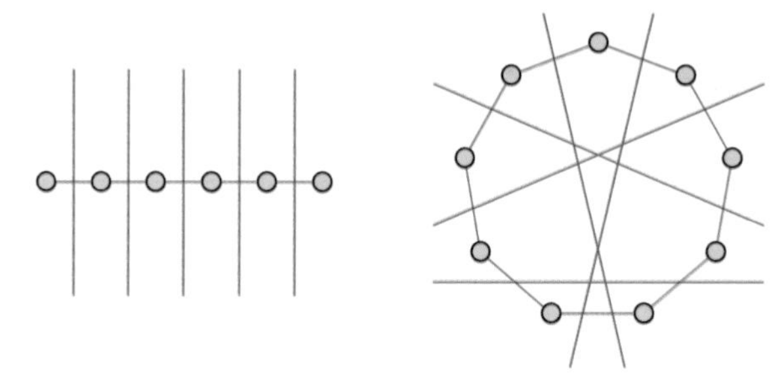

Figure 17.2. Illustration for Example 17.8. Shattering $\text{LINE}(n)$ requires each of $n-1$ consecutive pairs to be separated, for a total of $n-1$ lines (left). Shattering $\text{POLYGON}(n)$ requires each of n consecutive pairs to be separated, two per line, for a total of $\lceil n/2 \rceil$ lines (right).

Example 17.8

The line and polygon configurations have

$$\text{SHATTER}\big(\text{LINE}(n)\big) = n - 1$$

and

$$\text{SHATTER}\big(\text{POLYGON}(n)\big) = \left\lceil \frac{n}{2} \right\rceil,$$

respectively. For, the $n-1$ consecutive pairs of points in $\text{LINE}(n)$ each must be separated by a separate line, and each line can separate only two consecutive pairs of points in $\text{POLYGON}(n)$. It is straightforward to find sets of this many lines that shatter these configurations, by making sure that each consecutive pair of points is separated by exactly one separating line in the case of $\text{LINE}(n)$ or that at most one consecutive pair is split twice in the case of $\text{POLYGON}(n)$. See Figure 17.2 for an illustration.

Example 17.9

If S is a grid, it is shattered by the axis-parallel lines that separate its rows and columns (Figure 17.3). Therefore,

$$\text{SHATTER}\big(\text{GRID}(m, n)\big) \leq m + n - 2.$$

As Har-Peled and Jones (2017) observe, random point sets have significantly different shattering numbers than a grid. As they show, for mn randomly

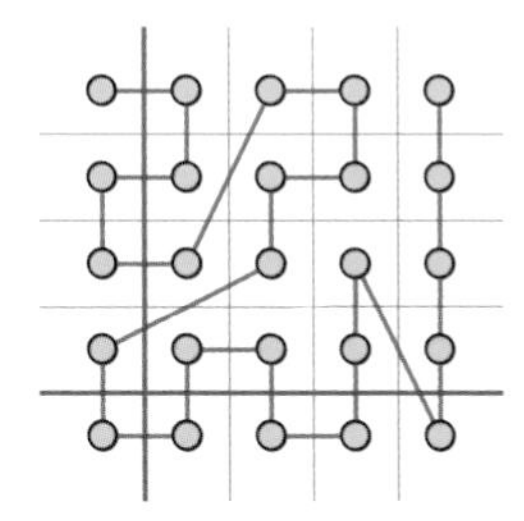

Figure 17.3. The configuration GRID$(5, 5)$ (yellow) is shattered by the eight axis-parallel lines separating the rows and columns. So, it has shattering number at most eight. Every spanning path (blue) has 24 edges, each of which is crossed by at least one of shattering lines. Therefore, at least one of these lines stabs at least $24/8 = 3$ edges. In this example, the two thicker red lines stab five edges each.

chosen points in a unit square, SHATTER is $\Theta((mn)^{2/3} \log^{O(1)} mn)$, significantly larger than Example 17.9.

Example 17.10

Instead of constructing an arrangement that shatters a given configuration, it is easier to construct a configuration that is shattered by a given arrangement. Given a number n, let A be an arbitrary arrangement of $\lceil \sqrt{2n} \rceil$ lines, no two of which are parallel and no three of which cross at the same point, so A has at least n cells. Form a configuration S by placing n points, at most one in each cell of A. Then SHATTER$(S) = \lceil \sqrt{2n} \rceil$, the minimum value possible for an n-point configuration.

Observation 17.11 (Welzl, 1992)

For every configuration S,

$$\text{PATH-STAB}(S) \geq \left\lceil \frac{|S| - 1}{\text{SHATTER}(S)} \right\rceil.$$

Proof. Let A be any arrangement that shatters S and has $|A| = \text{SHATTER}(S)$, where $|A|$ is the number of lines in A. Any spanning path of S has exactly $|S| - 1$ edges, and each edge of a spanning path is stabbed by at least one of the lines of A. Therefore, the average number of edges stabbed per line in A is at least $(|S| - 1)/|A|$. The result follows from the fact that the maximum number of edges stabbed by any line is an integer at least as large as this average. □

Corollary 17.12 (Welzl, 1992)

The configuration S constructed in Example 17.10 has

$$\text{PATH-STAB}(S) \geq \big(1 - o(1)\big)\sqrt{n/2}.$$

An alternative lower bound for the path-stabbing number follows from a connection between these combinatorial problems and the lengths of the segments in a spanning tree or path. Essentially, a long-enough collection of line segments inside a short-enough polygonal boundary must have high stabbing number, as the following lemma states more formally.

Lemma 17.13

Let E be a collection of line segments of total length ℓ, whose convex hull has perimeter h. Then the maximum stabbing number of E is at least $\lceil 2\ell/h \rceil$.

Proof. We use a standard tool from geometric measure theory, the existence of a measure on the space of lines in the plane with the property that, for any rectifiable curve C, the length of C equals the integral, over the space of lines with respect to this measure, of the number of intersection points of each line with C. When restricted to the lines that intersect the convex hull of E and normalized appropriately, this measure becomes a continuous probability distribution on lines such that the expected number of intersection points of each line with E is $2\ell/h$. The result follows from the facts that the supremum of a random variable is at least its expectation and that the maximum stabbing number is an integer. □

Example 17.14

Either the shattering method or the length method can be used to show that $\text{PATH-STAB}\big(\text{GRID}(n, n)\big)$ is $\Omega(n)$. For the shattering method, any spanning path of the grid has $n^2 - 1$ edges and the grid has a shattering set of $2(n-1)$ axis-parallel lines, so one of these lines must stab at least

$$\text{PATH-STAB}\big(\text{GRID}(n, n)\big) \geq \frac{n^2 - 1}{2(n-1)} = \frac{n+1}{2}$$

edges. (For instance, Figure 17.3 depicts this for $n = 5$.) For the length method, any spanning path has length at least $\ell = n^2 - 1$ and the convex hull

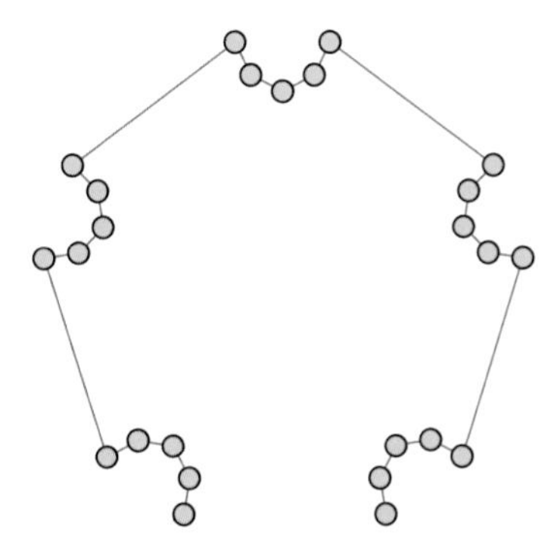

Figure 17.4. Construction of a family of configurations with bounded PATH-STAB(S) and unbounded CONVEX-PARTITION(S).

has perimeter $h = 4(n-1)$, so there must exist a line that stabs at least

$$\frac{2\ell}{h} = \frac{2(n^2-1)}{4(n-1)} = \frac{n+1}{2}$$

edges, the same bound.

On the other hand, there are sets of points for which only one of these two methods works well. The length method applies to randomly generated sets of n points in the unit square, showing that they have path stabbing number $\Omega(\sqrt{n})$, while the shattering method does not. The shattering method applies to distorted grids with nonuniform spacing between grid lines, while in general the length method does not.

17.3 Inequalities and Complexity

Observation 17.15

PATH-STAB $\ll$ WEAK-PARTITION.

Proof. A path with stabbing number $O(\text{WEAK-PARTITION}(S))$ may be found by partitioning S into WEAK-PARTITION(S) weakly convex subconfigurations, tracing around the convex hull of each subconfiguration to form a path, and connecting these paths into a single path in an arbitrary sequence.

To complete the proof, we must construct a family of configurations with stabbing number $O(1)$ and unbounded WEAK-PARTITION. This may be obtained by replacing each vertex of a regular n-gon by a concave chain of n vertices and connecting these chains into a path in the order given by the original n-gon (Figure 17.4). If the concave chains are sufficiently close to the polygon vertices they replace, any line can only stab two of the chains and two of the connecting edges, so PATH-STAB$(S) = O(1)$ for these configurations. However, any weakly convex subconfiguration must either remain within a single one of the concave chains or use $O(1)$ points per chain, so its size is $O(n)$. In order to cover all n^2 points by chains of this size, we must have WEAK-PARTITION$(S) = \Omega(n)$. □

Open Problem 17.16

Does PATH-STAB have finite obstacle size?

Several stabbing-number problems were proven NP-hard by Fekete et al. (2008), but they do not include PATH-STAB.

Theorem 17.17 (Har-Peled, 2009)

PATH-STAB *can be approximated to within a logarithmic approximation ratio in polynomial time.*

Open Problem 17.18

What is the computational complexity of PATH-STAB?

Open Problem 17.19

What relation, if any, exists between UNIVERSAL and PATH-STAB?

Observation 17.20

SHATTER $\doteq$ SIZE. *The shatter parameter has polynomial obstacle size.*

Proof. This follows from the fact that the shattering number can only range between $\sqrt{2n}$ (the minimum number of lines that can form n distinct arrangement cells, from Example 17.10) and $n-1$ (the shattering number of collinear points, from Example 17.8).[4] Turning this relation around, the obstacle size is at most quadratic in the shatter parameter: every configuration of size larger than quadratic in the parameter is forbidden, so the minimal forbidden configurations cannot be larger than this. □

Theorem 17.21 (Freimer, 2000)

Computing SHATTER(S) *is* NP*-hard, but it can be approximated with an approximation ratio of* $1 + \ln |S|$.

An alternative approximation algorithm, but still with a logarithmic approximation ratio, was given by Har-Peled and Jones (2017).

17.4 Unstabbed Segments

The parameter we define next plays a central role in many of our proofs of well-quasi-ordering, despite its somewhat artificial appearance.

Definition 17.22

Let S be any configuration. We define an *unstabbed segment* of S to be a line segment s with the following properties.

- The endpoints of s both belong to S.
- Segment s passes through at least one additional point of S, other than its endpoints.
- For every line L defined by a pair of points of $S \setminus s$, L does not stab s.

An unstabbed segment is *maximal* if it is not part of a longer unstabbed segment. If X is a set of unstabbed setments, we define a point of S to be a *free point* if it does not belong to any segments of X. We define the parameter UNSTABBED(S) to be the minimum, over sets X of unstabbed segments of X, of twice the size of X plus the number of free points of S in X.

The reason that we double the number of maximal unstabbed segments in the definition of UNSTABBED(S) is to ensure that UNSTABBED(S) is a monotone parameter. Removing a point from a maximal unstabbed segment that covers three points in S could cause the other two points to become free, and the

[4] Freimer (2000).

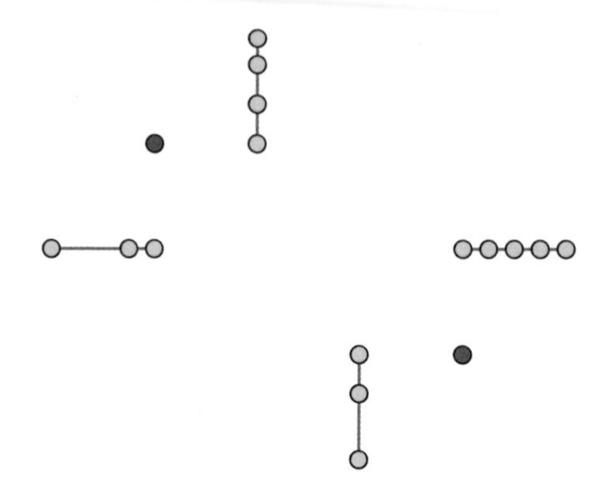

Figure 17.5. A configuration S with UNSTABBED$(S) = 10$. The four maximal unstabbed segments are shown in blue, and the two free points are shown in red.

doubled number of segments prevents the parameter value from increasing in this case. The reason we allow for a free choice of X rather than simply choosing X to be the set of all unstabbed segments is the same, to ensure that UNSTABBED(S) is monotone. If we required all unstabbed segments to belong to X, then deleting a point of S could create many new unstabbed segments (stabbed by lines involving the deleted point); however, in this case, we can still leave X unchanged, preventing the parameter from increasing. An example of this parameter is shown in Figure 17.5.

The segment from p to q is stabbed if and only if there exist points r and s, neither of which is collinear with p and q, such that the triples (p, r, s) and (q, r, s) have orientations with nonzero and opposite signs. Therefore, the definition of the parameter UNSTABBED(S) depends only on the configuration and not on the specific point set realizing the configuration.

Observation 17.23

LINE-COVER $\ll$ UNSTABBED $\ll$ OFFLINE.

Proof. If S can be covered by segments, with k segments plus free points, then it can also be covered by k lines, one per segment or free point. On the other hand, GRID$(3, n)$ has

$$\text{LINE-COVER}\big(\text{GRID}(3, n)\big) = 3$$

and

$$\text{UNSTABBED}\big(\text{GRID}(3, n)\big) = 3n,$$

because (for $n \geq 3$) it has no unstabbed segments: every axis-parallel segment is stabbed by a perpendicular line, and every non–axis-parallel segment is

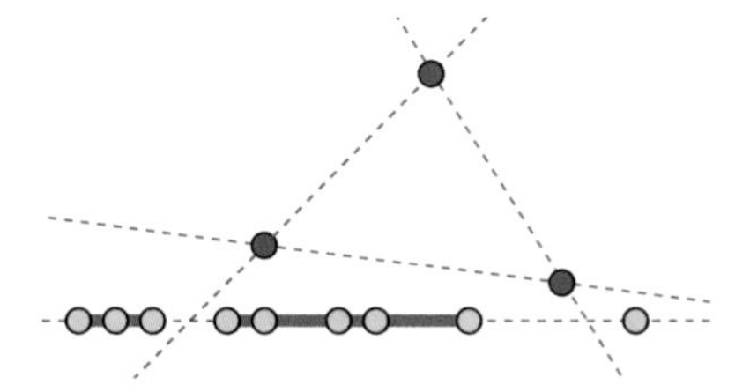

Figure 17.6. Constructing a small number of unstabbed segments for a configuration with a small value of OFFLINE(S) by cutting the heavy line (blue) into segments at each point where another line stabs it.

stabbed by the line through the other two corners of its axis-parallel bounding box. Therefore, LINE-COVER $\ll$ UNSTABBED.

If S has OFFLINE$(S) = k$, then we can partition the heavy line of S into $O(k^2)$ unstabbed segments by cutting it at each point where it is crossed by a line through two off-line points (Figure 17.6), so

$$\text{UNSTABBED}(S) = O\big(\text{OFFLINE}(S)^2\big).$$

On the other hand, GRID$(2, n)$ has

$$\text{UNSTABBED}\big(\text{GRID}(2, n)\big) = 4$$

and

$$\text{OFFLINE}\big(\text{GRID}(2, n)\big) = n.$$

Therefore, UNSTABBED $\ll$ OFFLINE. □

Observation 17.24

HEAVY-LINES $\ll$ UNSTABBED.

Proof. We partition the heavy lines of a configuration S into two subsets: the ones that include at least one unstabbed segment of S and the ones that do not. The number of heavy lines that include at least one unstabbed segment is at most the number of maximal unstabbed segments, which is at most UNSTABBED(S) as each unstabbed segment either directly contributes two units to UNSTABBED or contains an interior free point that does not belong to any other unstabbed segment and contributes one unit to unstabbed. The remaining heavy lines each contain at least three points that are either free or are endpoints of unstabbed segments. They are determined by any of the (at least three) pairs of these points. The number of points that are either free or are endpoints of unstabbed segments is at most UNSTABBED(S) (possibly smaller, if some segments share endpoints). Therefore, putting the bounds on these two kinds of heavy lines together, we have

$$\text{HEAVY-LINES}(S) \le \text{UNSTABBED}(S) + \binom{\text{UNSTABBED}(S)}{2} \Big/ 3.$$

To complete the proof, we must construct a family of configurations for which HEAVY-LINES(S) is bounded but UNSTABBED(S) is unbounded. To do so, consider any configurations that are in general position, so that HEAVY-LINES$(S) = 0$ but UNSTABBED$(S) = |S|$. □

17.5 Well-Quasi-Ordering

We show in this section that the configurations for which UNSTABBED(S) is bounded are very well-behaved but that they form a minuscule fraction of all configurations.

Lemma 17.25

Let S be a set of points and let C be any set of maximal unstabbed segments of S. Then no segment s of C can be stabbed by any line through two points p and q of $(S \cup C) \setminus s$.

Proof. Any such line L can be perturbed (maintaining its property of stabbing s) until it passes through two points p' and q' of S that are either p or q, points on the same segment of C as p or q, or points of $S \cap s$. If p' and q' are still both disjoint from s, we have stabbed s by a segment through two points of S. And if one of p' or q' belongs to s, we have stabbed a segment of C containing p or q by a segment through two points of S. So in all cases, we have found a contradiction to the assumption that all segments in C are unstabbed, proving that the initial line L cannot exist. □

Lemma 17.26

Let S be a set of points and let C be any set of maximal unstabbed segments of S. Define the position *of any point $p \in S \cup C$ to be the segment s of C such that p is interior to s, if such a segment exists, or to be p itself in the other cases (where p is a free point or endpoint of C). Then any two triples of points (p, q, r) and (p', q', r') with the same positions have the same orientations.*

Proof. If $p \neq p'$, then continuously slide p to p', then do the same for q and q' and for r and r', producing a continuous motion from (p, q, r) to (p', q', r') in which the positions remain invariant. The orientation of the triple can only change at a time during this motion where the three points become collinear, but by Lemma 17.25 a collinearity can only exist for three points whose positions are single points or that are constrained to remain collinear throughout the entire motion. □

Definition 17.27

Let S be a set of points and let C be any set of maximal unstabbed segments of S. Define the *cover order type* of the pair (S, C) to be the function mapping triples of positions to orientations, as described by Lemma 17.26.

For sets of k unstabbed segments, there can only be $O(k^3)$ triples, so the number of cover order types is bounded by a function of k.

Theorem 17.28

For any fixed k, the configurations S with $\text{UNSTABBED}(S) \leq k$ *are well-quasi-ordered. The number of such configurations with n points is bounded by a polynomial in n. Any monotone parameter is nonuniform fixed-parameter tractable on this subset of configurations.*

Proof. We can characterize a configuration by the cover order type of (S, C), where C is the subset of maximal unstabbed segments of S that determines $\text{UNSTABBED}(S)$, and by the number of interior points of S in each segment of C (Figure 17.7). There are $O(1)$ cover order types and $O(n^{\lfloor k/2-1 \rfloor})$ parameterizations of this type for configurations with n points, a polynomial bound.

There can be no infinite antichain of configurations S all of which have $\text{UNSTABBED}(S) \leq k$. Because there is only a finite number of cover order types, any such antichain would have to include an infinite subset all having the same cover order type. However, configurations with the same cover order type are distinguished only by their tuples of numbers of interior points on each segment, and these tuples are well-quasi-ordered by Dickson's lemma. Therefore, the configurations S with $\text{UNSTABBED}(S) \leq k$ are well-quasi-ordered.

To test whether the $\rho(S) < \ell$ for any monotone parameter ρ, we precompute a list, for each possible cover order type, of the parameterizations of the obstacles for $\rho(S) < \ell$ among the configurations S with that order type. These lists are all finite by well-quasi-ordering. Then, for any given S, we look up the

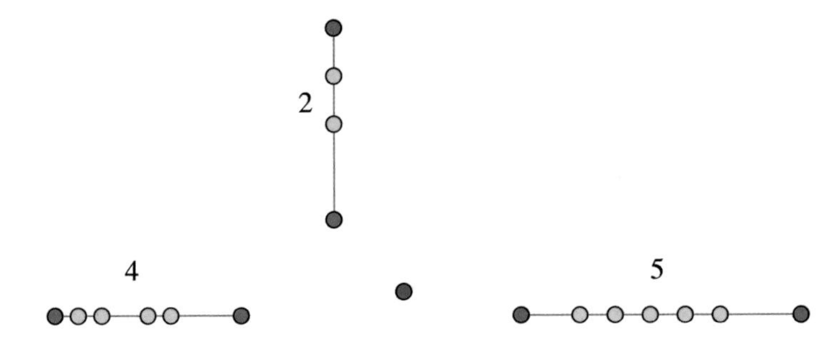

Figure 17.7. Parameterizing a configuration with a given cover order type (determined by the positions of the blue unstabbed segments and red free point) by the tuple of numbers (2, 4, 5) of interior points on each unstabbed segment.

list for its cover order type and test for each obstacle whether the number of interior points of S on each segment dominates the number for the obstacle. It takes polynomial time to find the maximal unstabbed segments of S, an amount of time bounded by a function of the parameter to select the subset of segments determining UNSTABBED(S) and compute its cover order type, and $O(k)$ to test each of $O(1)$ obstacles, so it is nonuniform fixed-parameter tractable. □

17.6 Applications of Unstabbed Segments

The purpose of introducing the UNSTABBED parameter was to use it to prove that other more natural classes of configurations are well-quasi-ordered. We do so here.

Proof of Theorem 8.13. Recall that this theorem states that the configurations with OFFLINE(S) $< k$ are well-quasi-ordered. This follows immediately from Theorem 17.28 and Observation 17.23, according to which UNSTABBED $\ll$ OFFLINE. □

Proof of Theorem 8.16, part I. The first part of this theorem states that the configurations with LINE-COVER(S) ≤ 2 are well-quasi-ordered. For such a configuration, consider a set of (at most two) lines that cover S and partition these lines into four rays at the point where they cross (if there is such a point; see Figure 17.8, left). Each ray produces either a single maximal unstabbed segment or up to two free points, in either case contributing at most two to the total value of UNSTABBED(S), so every configuration S with LINE-COVER(S) ≤ 2 has UNSTABBED(S) ≤ 8 (Figure 17.8). The result follows from Theorem 17.28. □

Proof of Theorem 9.7, part I. The first part of Theorem 9.7 concerns the configurations for which MAX-GENERAL(S) < 5 and states that the number of

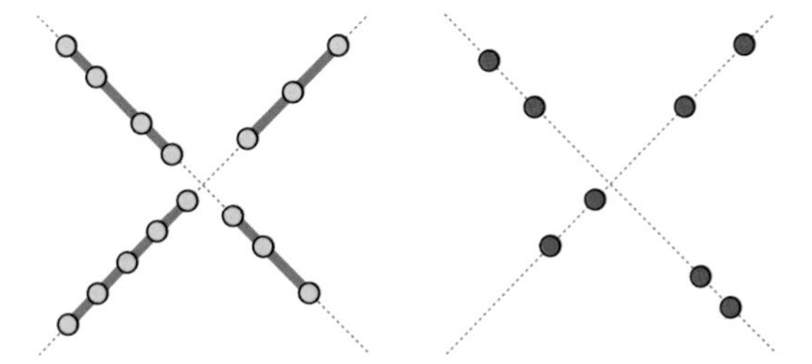

Figure 17.8. Illustration for the proof of Theorem 8.16. Any configuration S with LINE-COVER$(S) = 2$ can be partitioned into four subsets by partitioning each line into two rays that meet at the crossing points of the lines. Regardless of whether each subset forms a single maximal unstabbed segment (left) or up to two free points (right), UNSTABBED(S) will be at most eight.

these configurations is polynomially bounded and that they are well-quasi-ordered. This follows from the first part of Theorem 11.22 (proof below), which states the corresponding results for MAX-CONVEX, as MAX-CONVEX$(S) <$ MAX-GENERAL(S). □

Next, we wish to prove the first part of Theorem 11.22 concerning the configurations for which MAX-CONVEX$(S) < 5$. Rather than doing so directly, it is convenient to have the following lemma, which encapsulates most of the work of the proof.

Lemma 17.29

There is a positive integer q with the following property. Suppose that some configuration S has a line L, such that at least q disjoint segments between points of S on L are stabbed by lines through two points of S. Then S contains a convex pentagon.

Proof. We go through a sequence of steps in which we impose additional structure on S, L, and the stabbing lines of the segments of L, until we can eventually prove that the desired convex pentagon exists.

- The known relation MAX-CONVEX $\doteq$ LINE-COVER implies that the configurations S with MAX-CONVEX$(S) < 5$ can be covered by at most ℓ lines for some positive integer ℓ. We may assume, therefore, that the given configuration S in the statement of the lemma can be covered by at most ℓ lines (not necessarily including L), for otherwise the conclusion of the lemma would already follow.
- We may classify the lines that stab segments of L into $O(\ell^2)$ types, where a type describes the identity of the two covering lines that contain the two defining points of the stabbing line, the order in which the stabbing line

passes through each of these three lines, and the order in which the stabbing line passes through the two sides of each of these three lines. (A stabbing line may belong to more than one of these types, if it has more than one pair of defining points or a point on more than one covering line.)

- Because there are only $O(\ell^2)$ types of stabbing lines, there is a subset of $q' = \Omega(q/\ell^2)$ of the stabbing lines that are all of the same type. These lines are all determined by points on two particular covering lines of S, and they all cross these two covering lines and L in the same order. We may restrict S to the subconfiguration S' of points on L and on these two covering lines, and replace q by q'. If we can ensure that restricted problems of this form always contain a convex pentagon, for sufficiently large values of q', then so will the original configuration, for sufficiently large values of q (larger than the threshold value for q' by a factor of $O(\ell^2)$).
- Sort the remaining stabbing lines by the position along L of the segments that they stab, and choose at most one stabbing line per segment. The sorted order along L may differ from the order in which these stabbing lines cross each of the two covering lines.
- We apply the Erdős–Szekeres theorem (Theorem 14.13) twice, to the sequences of crossings of the stabbing lines with the two covering lines. By doing so, we obtain a subsequence of $q'' = \Omega(q'^{1/4})$ stabbing lines that are consistently ordered along all three of L and the two covering lines.
- At a further reduction of q'' by a factor of two, we may ensure that (even if each of the stabbing lines passes through a point of S' on L instead of stabbing a segment between two points) there exists at least one point of S' between each two consecutive points where S' is crossed by the subsequence of ordered stabbing lines, and at least one point of S' on the two rays of L outside the part of L that is crossed by the stabbing lines.
- We restrict the set of stabbing lines to its ordered subsequence. We restrict S' to a subconfiguration in which each segment or ray of L between the crossing points of the stabbing lines contains exactly one point, and in which the only points off L are the ones that define these stabbing lines. And we reduce q' to q''. If we can ensure that restricted problems of this form always contain a convex pentagon, for sufficiently large values of q'', then so will the original configuration, for sufficiently large values of q (larger than the square of the threshold value for q'' by a factor of $O(\ell^2)$).

This analysis reduces the problem to a finite set of cases, according to how L and the two remaining covering lines are positioned and according to whether the ordering of the stabbing lines along each covering line is consistent with the ordering along L, reversed from the ordering along L, or constant. These cases are depicted in Figure 17.9. As the figure shows, we take $q'' = 3$; that is, at most three stabbing lines are needed in each case. The cases are the following.

- The three stabbing lines might all pass through a single point of one of the two covering lines, when the application of the Erdős–Szekeres theorem gives us a constant subsequence. In this case, there are three subcases,

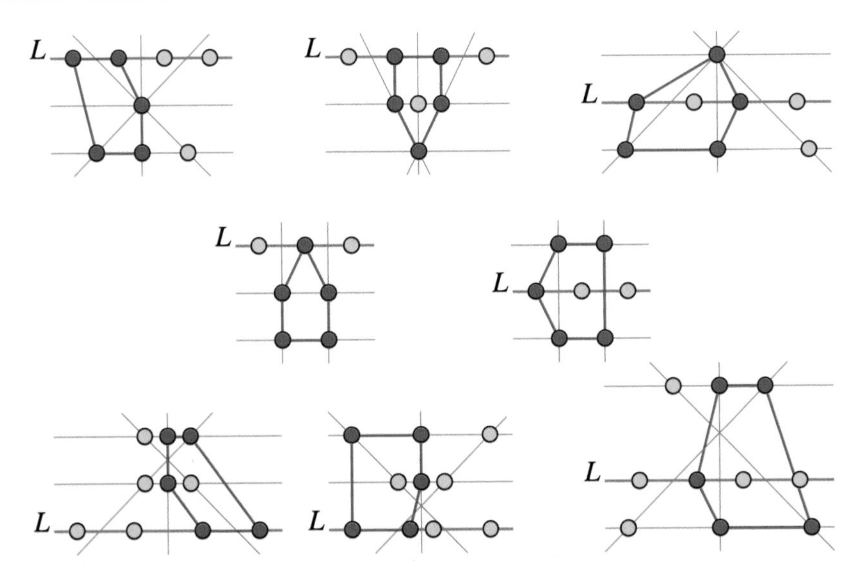

Figure 17.9. Cases for Lemma 17.29. These show the different ways that a line L can be stabbed by three lines through points on two other covering lines, with the three stabbing lines crossing L and the two covering lines in a consistent order, and with one point of L on each of the segments formed by the stabbing lines. In the top three cases the stabbing lines all pass through a single point on one of the covering lines. In the middle two cases the stabbing lines cross L and the two covering lines in the same order. In the bottom cases one of these crossing orders is reversed. Each case necessarily includes a convex pentagon (red).

according to whether the middle line among L and the two covering lines is L, the covering line with the single point, or the other covering line. These are the three top cases of Figure 17.9.

- The three stabbing lines might all pass through L and the two covering lines in the same order, so that their crossings (if they are not parallel) are outside the configuration. In this case, there are two subcases: L might be between the two covering lines, or the middle line might be one of the two covering lines. These are the two middle cases of Figure 17.9. They require only two stabbing lines.
- The order in which the three stabbing lines cross L and the two covering lines might be reversed along one line from the other two, so that the stabbing lines all cross between two of the lines. Again we have three subcases: L might be the middle of the three lines, it might be the outside line that has crossings between it and the middle covering line, or it might be the other outside line. These are the three lower cases of Figure 17.9. For this case, the three stabbing lines do not necessarily all cross at a single point, although they are drawn that way in the figure.

As can be seen from the figure, we can identify a convex pentagon in each of the cases. We need to ensure, however, that in each case the pentagon will

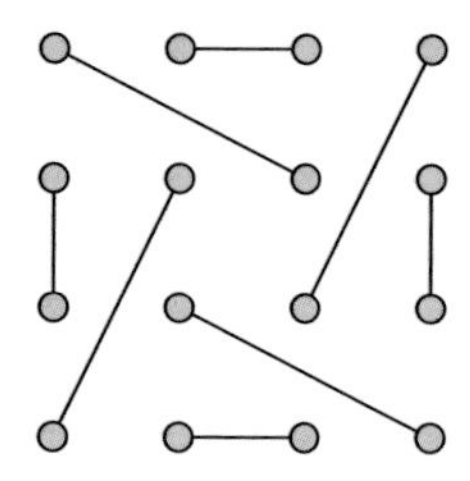

Figure 17.10. Solution to the puzzle at the beginning of the chapter: eight disjoint line segments in GRID(4, 4) that block all lines of sight through the grid.

necessarily exist in any configuration described by that case, and not just for the one in the drawing. To make sure that this is true, one can verify that, for each triple of consecutive points in each pentagon of each case (40 triples in total) there is a line of the configuration (L, one of the stabbing lines, or one of the covering lines) that passes through two of the points in the triple. Because of the way these cases were constructed, the third point of the triple can only be on one side of this line, guaranteeing that the orientation of the triple is what we want it to be. We can also verify that no two pentagon edges can cross, because all pairs of nonconsecutive edges are separated by a line of the construction (including lines containing one of the edges) with one exception: the case at the bottom center of the figure. In that case, the two nonseparated edges still cannot cross, because if they did they would cause their endpoints on the bottom line to be out of order. Thus, all cases have a non–self-crossing pentagon with all interior angles convex, which must be a convex pentagon. □

Proof of Theorem 11.22, part I. The first part of Theorem 11.22 concerns the configurations for which MAX-CONVEX$(S) < 5$ and states that the number of these configurations is polynomially bounded and that they are well-quasi-ordered. To prove this, we show that these configurations have bounded values of UNSTABBED(S).

Because of the relation MAX-CONVEX $\doteq$ LINE-COVER, any configuration S with MAX-CONVEX < 5 would necessarily have to have a bounded number of covering lines. To be specific, let ℓ denote the largest possible number of covering lines for any configuration S with this property. Additionally, let q be the number given by Lemma 17.29. Then each of the covering lines of S can be divided into $O(q)$ unstabbed segments and free points. It follows that UNSTABBED$(S) = O(q\ell) = O(1)$. □

18 The Big Picture

As we have seen, many long-known problems from discrete geometry can be brought into the picture of monotone properties and parameters of configurations. Along the way, we have found obstacles and algorithms for many of these parameters. We have also shown that many pairs of these parameters are either equivalent for the purposes of fixed-parameter tractability (when each is bounded by a function of the other) or related by bounds in only one direction. Figure 18.1 shows some of these parameters and their relations.

In the next two sections, we summarize the key results about these properties and parameters.

18.1 Properties

COLLINEAR is the property of having all points on a single line (Definition 8.2). It may be tested in linear time (Section 8.4). There is only one collinear configuration of each size n, the configuration LINE(n). It has one obstacle, POLYGON(3).

CONVEX is the property of forming the vertices of a convex polygon (Definition 11.1). It may be tested in polynomial time by computing the convex hull (Section 11.4). There is only one convex configuration of each size n, the configuration POLYGON(n). It has a finite set of obstacles.

GENERAL-POSITION is the property of being in general position, true of point sets with no three in a line (Definition 9.1). Thus, it has one obstacle, LINE(3). It may be tested in quadratic time (Section 9.2).

INTEGER-COORDINATES(S) is the property of being realizable with integer coordinates, true of subconfigurations of grids (Definition 13.1). It includes all

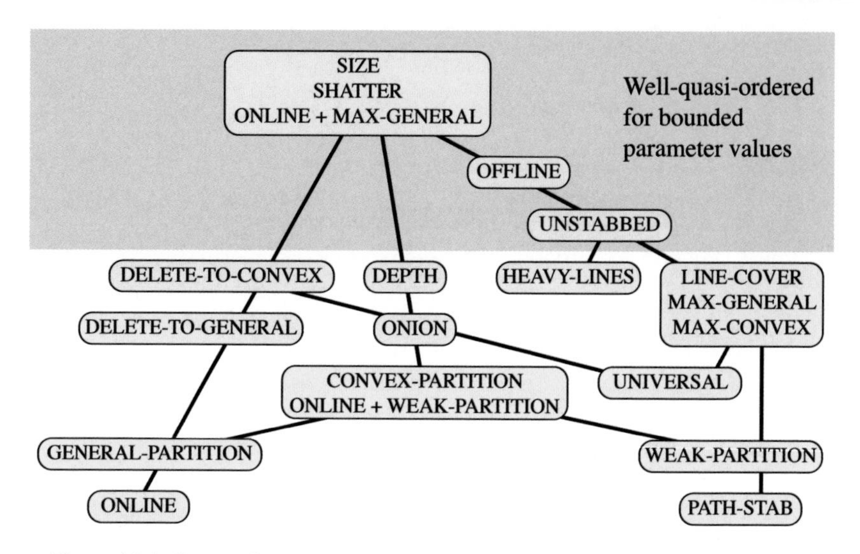

Figure 18.1. Some of the parameters we have defined. Parameters within the same shaded box are equivalent ($\doteq$). A dark line from one parameter to another indicates that the lower of the two parameters in the illustration is upper bounded by a function of the upper one ($\ll$). For the parameters in the upper blue region, the configurations with a bounded parameter value are well-quasi-ordered; for the remaining parameters, this is not the case.

configurations in general position and all weakly convex configurations. We do not know its obstacles or computational complexity (Open Problem 13.4).

INTEGER-DISTANCES(S) is the property of being realizable with all pairwise distances equal to integers (Definition 13.1). It includes all configurations in convex position (Section 13.2). We do not know its obstacles or computational complexity (Open Problem 13.4).

STRETCHED is a property that characterizes the configurations generated by the stretching transformation (Definition 14.5), used to represent permutations by point sets (Observation 14.7). It may be tested in polynomial time (Theorem 14.23). Finding subconfigurations of stretched configurations is fixed-parameter tractable (Theorem 14.26).

WEAKLY-CONVEX is the property of being in weakly convex position (Definition 12.1). It has a finite set of obstacles and may be computed in polynomial time (Section 12.2). The weakly convex configurations are not well-quasi-ordered (Theorem 12.3).

18.2 Parameters

CONVEX-PARTITION is the fewest number of convex polygons whose vertices cover all points of the configuration (Definition 11.1). It is NP-hard, but can be tested in polynomial time for any constant parameter value (Theorem 11.15), and has a logarithmic approximation algorithm (Observation 11.17). It has infinitely many obstacles (Theorem 11.11).

DELETE-TO-CONVEX is the fewest points that must be deleted to make the rest be in convex position (Definition 11.1). It has polynomial obstacle size (Section 11.3) and may be tested in polynomial time (Theorem 11.14).

DELETE-TO-GENERAL is the fewest points that must be deleted to make the rest be in general position (Definition 9.1). It has polynomial obstacle size, is NP-hard and APX-hard (Theorem 9.3), and is fixed-parameter tractable. It can approximated to within a factor of three (Theorem 9.18).

DEPTH is the largest number d such that, for some point p of a configuration, every closed halfplane containing p contains at least d points of the configuration (Definition 12.15). It has polynomial obstacle size (Observation 12.16) and can be computed in polynomial time (Theorem 12.20).

GENERAL-PARTITION is the fewest number of subconfigurations in a partition into general-position subconfigurations (Definition 9.1). It has infinitely many obstacles, and testing whether it is below a given threshold is NP-complete, even when that threshold is three (Theorem 10.16). It can be approximated to within a factor of $O(\sqrt{|S|})$ (Theorem 10.20).

HEAVY-LINES is the number of heavy lines in a configuration. Its maximum values are the subject of the orchard-planting problem (Section 8.1).

IN-TRIANGLE is the largest number of points interior to any triangle of a configuration. It may be computed in polynomial time, and when it is large the configuration necessarily also includes a large empty polygon (Section 12.6).

LINE-COVER is the smallest number of lines needed to cover all points of a configuration (Definition 8.2). Testing whether it is below a given threshold is NP-complete (Theorem 8.9) but fixed-parameter tractable (Theorem 8.10). Approximating it is APX-hard, but it has a logarithmic approximation algorithm (Theorem 8.17).

MAX-CONVEX is the size of the largest convex subconfiguration (Definition 11.1). It has polynomial obstacle size (Section 11.3) and may be tested in polynomial time (Theorem 11.14).

MAX-GENERAL is the size of the largest general-position subconfiguration (Definition 9.1). It has polynomial obstacle size, is NP-hard and APX-hard (Theorem 9.3), and is fixed-parameter tractable (Theorem 9.5). It can be approximated to within a factor of $O(\sqrt{|S|})$ (Theorem 9.16).

OFFLINE is the fewest points to delete so that the rest become collinear (Definition 8.2). It may be computed in quadratic time (Section 8.4) and has polynomial obstacle size (Section 8.7).

ONION is the number of nested convex polygons into which a configuration can be decomposed (Definition 12.6). It can be computed in polynomial time (Theorem 12.8) and has finite obstacle size (Theorem 12.9).

ONLINE is the most points on any single line of a configuration (Definition 8.2). It may be computed in quadratic time (Section 8.4) and has polynomial obstacle size (Section 8.7).

PATH-STAB is the number of crossings between a path through all points of a configuration and a line, with the path chosen to minimize this crossing number and the line chosen as the one that crosses this path as many times as possible (Definition 17.2).

SHATTER is the fewest lines needed to separate all pairs of points in a configuration (Definition 17.7). It has polynomial obstacle size (Observation 17.20) and is NP-hard, but logarithmically approximable (Theorem 17.21).

SIZE is the number of points in a configuration (Definition 3.6). It has bounded obstacles (Example 5.11) and is trivially computable in polynomial time.

UNIVERSAL is a parameter giving the largest n such that a configuration can be used to draw all n-vertex planar graphs (Definition 16.1). It has finite obstacle size (Observation 16.8).

UNSTABBED measures the complexity of a decomposition of a configuration into unstabbed segments and free points (Definition 17.22). The configurations for which UNSTABBED is bounded are well-quasi-ordered (Theorem 17.28).

WEAK-PARTITION is the fewest number of weakly convex subconfigurations whose points cover the configuration (Definition 12.1). It has infinitely many obstacles. Testing whether it is at most k, when k is constant, can be done in polynomial time, and it has a logarithmic approximation algorithm (Section 12.2).

18.3 Only the Beginning

Although some of these topics date back to the early eighteenth century (for the orchard-planting problem) or even earlier (for integer distances), the general study of forbidden configurations seems ripe for much additional research. Several of its problems, including orchard-planting and the happy ending problem, have seen significant progress in the last few years. And looking at these topics from the point of view of their obstacles and obstacle-based algorithmics has raised many more questions, not previously addressed in the literature. Although we have already answered some of these questions in these pages, we have left many more for future research.

Bibliography

Abu-Khzam, Faisal N. 2010. A kernelization algorithm for d-hitting set. *J. Comput. System Sci.*, **76**(7), 524–531. (Cited on page 51.)

Agarwal, Pankaj K. 1992. Ray shooting and other applications of spanning trees with low stabbing number. *SIAM J. Comput.*, **21**(3), 540–570. (Cited on page 191.)

Agarwal, Pankaj K., and Procopiuc, Cecilia M. 2003. Approximation algorithms for projective clustering. *J. Algorithms*, **46**(2), 115–139. (Cited on page 61.)

Aichholzer, Oswin, Aurenhammer, Franz, and Krasser, Hannes. 2002. Enumerating order types for small point sets with applications. *Order*, **19**(3), 265–281. (Cited on page 10.)

Aichholzer, Oswin, Huemer, Clemens, Kappes, Sarah, Speckmann, Bettina, and Tóth, Csaba D. 2007. Decompositions, partitions, and coverings with convex polygons and pseudo-triangles. *Graphs Combin.*, **23**(5), 481–507. (Cited on page 132.)

Aichholzer, Oswin, Miltzow, Tillmann, and Pilz, Alexander. 2013. Extreme point and halving edge search in abstract order types. *Comput. Geom. Th. & Appl.*, **46**(8), 970–978. (Cited on page 34.)

Ailon, Nir, and Chazelle, Bernard. 2005. Lower bounds for linear degeneracy testing. *J. ACM*, **52**(2), 157–171. (Cited on page 59.)

Alimonti, Paola, and Kann, Viggo. 2000. Some APX-completeness results for cubic graphs. *Theoret. Comput. Sci.*, **237**(1–2), 123–134. (Cited on page 45.)

Alon, Noga. 2012. A non-linear lower bound for planar epsilon-nets. *Discrete Comput. Geom.*, **47**(2), 235–244. (Cited on page 71.)

Anderson, David Brent. 1979. Update on the no-three-in-line problem. *J. Combin. Theory Ser. A*, **27**(3), 365–366. (Cited on page 73.)

Anning, Norman H., and Erdős, Paul. 1945. Integral distances. *Bull. Amer. Math. Soc.*, **51**(8), 598–600. (Cited on page 142.)

Appel, Kenneth, and Haken, Wolfgang. 1989. *Every Planar Map Is Four Colorable.* Contemporary Mathematics, vol. 98. Providence, RI: American Mathematical Society. (Cited on page 45.)

Arkin, Esther M., Fekete, Sándor P., Hurtado, Ferran, Mitchell, Joseph S. B., Noy, Marc, Sacristán, Vera, and Sethia, Saurabh. 2003. On the reflexivity of point sets. Pages 139–156 of: *Discrete and Computational Geometry: The*

Goodman–Pollack Festschrift. Algorithms and Combinatorics, vol. 25. Berlin: Springer. (Cited on pages 106, 117, 118, and 131.)

Aronov, Boris, Erdős, Paul, Goddard, Wayne D., Kleitman, Daniel J., Klugerman, Michael, Pach, János, and Schulman, Leonard J. 1994. Crossing families. *Combinatorica*, **14**(2), 127–134. (Cited on page 128.)

Balko, Martin, Kynčl, Jan, Langerman, Stefan, and Pilz, Alexander. 2017. Induced Ramsey-type results and binary predicates for point sets. *Electronic Journal on Combinatorics*, **24**(4), Paper #P4.24. (Cited on page 10.)

Balogh, Jozsef, and Solymosi, József. 2017. *On the number of points in general position in the plane*. Electronic preprint https://arxiv.org/abs/1704.05089. (Cited on page 93.)

Bannister, Michael J., Cheng, Zhanpeng, Devanny, William E., and Eppstein, David. 2014. Superpatterns and universal point sets. *J. Graph Algorithms Appl.*, **18**(2), 177–209. (Cited on pages 151, 181, and 183.)

Bannister, Michael J., Devanny, William E., Dujmović, Vida, Eppstein, David, and Wood, David R. 2016. Track layout is hard. Pages 499–510 of: *24th Int. Symp. Graph Drawing and Network Visualization (GD 2016)*. Lecture Notes in Computer Science, vol. 9801. Berlin: Springer. (Cited on page 180.)

Baran, Ilya, Demaine, Erik D., and Pătraşcu, Mihai. 2008. Subquadratic algorithms for 3SUM. *Algorithmica*, **50**(4), 584–596. (Cited on page 60.)

Barnett, Vic. 1976. The ordering of multivariate data. *J. Roy. Statist. Soc. Ser. A*, **139**(3), 318–355. (Cited on page 129.)

Beck, József. 2008. *Combinatorial Games: Tic-Tac-Toe Theory*. Encyclopedia of Mathematics and Its Applications, vol. 114. Cambridge: Cambridge University Press. (Cited on page 89.)

Björner, Anders, Las Vergnas, Michel, Sturmfels, Bernd, White, Neil, and Ziegler, Günter M. 1999. *Oriented Matroids*. 2nd edn. Encyclopedia of Mathematics and Its Applications, vol. 46. Cambridge: Cambridge University Press. (Cited on page 11.)

Borwein, Peter, and Moser, William O. J. 1990. A survey of Sylvester's problem and its generalizations. *Aequationes Mathematicae*, **40**(1), 111–135. (Cited on page 53.)

Brandenburg, Franz J. 2008. Drawing planar graphs on $\frac{8}{9}n^2$ area. Pages 37–40 of: *The International Conference on Topological and Geometric Graph Theory*. Electronic Notes in Discrete Mathematics, vol. 31. Amsterdam: Elsevier. (Cited on pages 181 and 182.)

Bremner, David, Chen, Dan, Iacono, John, Langerman, Stefan, and Morin, Pat. 2008. Output-sensitive algorithms for Tukey depth and related problems. *Stat. Comput.*, **18**(3), 259–266. (Cited on page 136.)

Brinkmann, Gunnar, and McKay, Brendan D. 2007. Fast generation of planar graphs. *MATCH Commun. Math. Comput. Chem.*, **58**(2), 323–357. (Cited on page 184.)

Bukh, Boris, Matoušek, Jiří, and Nivasch, Gabriel. 2011. Lower bounds for weak epsilon-nets and stair-convexity. *Israel J. Math.*, **182**, 199–208. (Cited on pages 151 and 157.)

Burr, Michael A., Rafalin, Eynat, and Souvaine, Diane L. 2004. Simplicial depth: An improved definition, analysis, and efficiency for the finite sample case. Pages 136–139 of: *16th Canadian Conference on Computational*

Geometry (CCCG '04), www.cccg.ca/proceedings/2004/58.pdf. (Cited on page 140.)

Burr, Stefan A., Grünbaum, Branko, and Sloane, Neil J. A. 1974. The orchard problem. *Geometriae Dedicata*, **2**, 397–424. (Cited on pages 54 and 55.)

Cabello, Sergio. 2006. Planar embeddability of the vertices of a graph using a fixed point set is NP-hard. *J. Graph Algorithms Appl.*, **10**(2), 353–363. (Cited on page 184.)

Carathéodory, Constantin. 1907. Über den Variabilitätsbereich der Koeffizienten von Potenzreihen, die gegebene Werte nicht annehmen. *Math. Ann.*, **64**(1), 95–115. (Cited on page 106.)

Cardinal, Jean, Hoffmann, Michael, and Kusters, Vincent. 2015. On universal point sets for planar graphs. *J. Graph Algorithms Appl.*, **19**(1), 529–547. (Cited on page 182.)

Chan, Timothy M. 2004. An optimal randomized algorithm for maximum Tukey depth. Pages 430–436 of: *15th ACM/SIAM Symposium on Discrete Algorithms (SODA '04)*. Philadelphia: Society for Industrial and Applied Mathematics. (Cited on page 137.)

Chan, Timothy M., and Lewenstein, Moshe. 2015. Clustered integer 3SUM via additive combinatorics. Pages 31–40 of: *47th Symposium on Theory of Computing (STOC '15)*. New York: Association for Computing Machinery. (Cited on page 60.)

Chazelle, Bernard. 1985. On the convex layers of a planar set. *IEEE Trans. Inform. Theory*, **31**(4), 509–517. (Cited on page 130.)

Chazelle, Bernard, and Welzl, Emo. 1989. Quasi-optimal range searching in spaces of finite VC-dimension. *Discrete Comput. Geom.*, **4**(5), 467–489. (Cited on page 191.)

Chazelle, Bernard, Guibas, Leonidas J., and Lee, Der-Tsai. 1985. The power of geometric duality. *BIT*, **25**(1), 76–90. (Cited on page 129.)

Chen, Jianer, Huang, Xiuzhen, Kanj, Iyad A., and Xia, Ge. 2006. Strong computational lower bounds via parameterized complexity. *J. Comput. System Sci.*, **72**(8), 1346–1367. (Cited on page 171.)

Chew, L. Paul, Goodrich, Michael T., Huttenlocher, Daniel P., Kedem, Klara, Kleinberg, Jon M., and Kravets, Dina. 1997. Geometric pattern matching under Euclidean motion. *Comput. Geom. Th. & Appl.*, **7**(1–2), 113–124. (Cited on page 36.)

Chor, Benny, and Sudan, Madhu. 1998. A geometric approach to betweenness. *SIAM J. Discrete Math.*, **11**(4), 511–523. (Cited on page 15.)

Chrobak, Marek, and Karloff, Howard. 1989. A lower bound on the size of universal sets for planar graphs. *SIGACT News*, **20**, 83–86. (Cited on page 181.)

Chvátal, Václav, and Klincsek, Gheza T. 1980. Finding largest convex subsets. Pages 453–460 of: *Proceedings of the Eleventh Southeastern Conference on Combinatorics, Graph Theory and Computing (Florida Atlantic Univ., Boca Raton, Fla., 1980)*, Vol. 2. Congressus Numerantium, vol. 29. (Cited on page 115.)

Clarkson, Kenneth L., and Varadarajan, Kasturi. 2007. Improved approximation algorithms for geometric set cover. *Discrete Comput. Geom.*, **37**(1), 43–58. (Cited on page 70.)

Craggs, D., and Hughes-Jones, Richard. 1976. On the no-three-in-line problem. *J. Combin. Theory Ser. A*, **20**(3), 363–364. (Cited on page 73.)

Croot, Ernie, Lev, Vsevolod, and Pach, Peter. 2016. *Progression-free sets in* $\mathbb{Z}_4^n$ *are exponentially small.* Electronic preprint arxiv:1605.01506. (Cited on page 92.)

Csima, Joseph, and Sawyer, Eric T. 1993. There exist $6n/13$ ordinary points. *Discrete Comput. Geom.*, **9**, 187–202. (Cited on page 55.)

de Fraysseix, Hubert, Pach, János, and Pollack, Richard. 1988. Small sets supporting Fary embeddings of planar graphs. Pages 426–433 of: *20th Symposium on Theory of Computing (STOC '88).* New York: Association for Computing Machinery. (Cited on pages 181 and 182.)

Dey, Tamal K. 1998. Improved bounds for planar k-sets and related problems. *Discrete Comput. Geom.*, **19**, 373–382. (Cited on page 18.)

Dickson, Leonard Eugene. 1913. Finiteness of the odd perfect and primitive abundant numbers with n distinct prime factors. *Amer. J. Math.*, **35**(4), 413–422. (Cited on page 23.)

Dolev, Danny, Leighton, F. Thomson, and Trickey, Howard. 1984. Planar embedding of planar graphs. *Advances in Computing Research*, **2**, 147–161. (Cited on page 182.)

Donoho, David L., and Gasko, Miriam. 1992. Breakdown properties of location estimates based on halfspace depth and projected outlyingness. *Ann. Statist.*, **20**(4), 1803–1827. (Cited on page 136.)

Downey, Rodney G., and Fellows, Michael R. 2013. *Fundamentals of Parameterized Complexity.* Texts in Computer Science. Berlin: Springer. (Cited on pages 46 and 47.)

Dudeney, Henry. 1917. 317. A puzzle with pawns. Page 94 of: *Amusements in Mathematics.* Edinburgh: Nelson. (Cited on page 73.)

Dumitrescu, Adrian, and Jiang, Minghui. 2015. On the approximability of covering points by lines and related problems. *Comput. Geom. Th. & Appl.*, **48**(9), 703–717. (Cited on pages 67 and 71.)

Eddy, William F. 1982. Convex hull peeling. Pages 42–47 of: *COMPSTAT 1982 5th Symposium held at Toulouse 1982.* Heidelberg: Physica-Verlag. (Cited on page 129.)

Edel, Yves. 2004. Extensions of generalized product caps. *Designs, Codes and Cryptography*, **31**(1), 5–14. (Cited on page 92.)

Edelsbrunner, Herbert, and Guibas, Leonidas J. 1989. Topologically sweeping an arrangement. *J. Comput. System Sci.*, **38**(1), 165–194. (Cited on pages 58 and 116.)

Edelsbrunner, Herbert, and Mücke, Ernst Peter. 1988. Simulation of simplicity: A technique to cope with degenerate cases in geometric algorithms. Pages 118–133 of: *4th Symposium on Computational Geometry (SoCG '88).* New York: Association for Computing Machinery. (Cited on page 11.)

Ellenberg, Jordan S., and Gijswijt, Dion. 2017. *On large subsets of* $\mathbb{F}_q^n$ *with no three-term arithmetic progressions. Ann. of Math.*, **185**(1), 339–343. (Cited on page 92.)

Eppstein, David. 2010. Happy endings for flip graphs. *J. Comput. Geom.*, **1**(1), 3–28. (Cited on page 110.)

Eppstein, David. 2014. Drawing arrangement graphs in small grids, or how to play planarity. *J. Graph Algorithms Appl.*, **18**(2), 211–231. (Cited on page 178.)

Eppstein, David, Overmars, Mark H., Rote, Günter, and Woeginger, Gerhard. 1992. Finding minimum area k-gons. *Discrete Comput. Geom.*, **7**(1), 45–58. (Cited on page 134.)

Erdős, Paul. 1986. On some metric and combinatorial geometric problems. *Discrete Math.*, **60**, 147–153. (Cited on page 80.)

Erdős, Paul. 1988. Some old and new problems in combinatorial geometry. Pages 32–37 of: *Applications of Discrete Mathematics (Clemson, SC, 1986)*. Philadelphia: Society for Industrial and Applied Mathematics. (Cited on page 80.)

Erdős, Paul, and Lovász, László. 1975. Problems and results on 3-chromatic hypergraphs and some related questions. Pages 609–627 of: Hajnal, András, Rado, Richard, and Sós, Vera T. (eds.), *Infinite and Finite Sets (to Paul Erdős on His 60th Birthday)*, vol. II. Amsterdam: North-Holland. (Cited on page 81.)

Erdős, Paul, and Rado, Richard. 1960. Intersection theorems for systems of sets. *J. London Math. Soc., 2nd Ser.*, **35**(1), 85–90. (Cited on page 49.)

Erdős, Paul, and Szekeres, George. 1935. A combinatorial problem in geometry. *Compositio Math.*, **2**, 463–470. (Cited on pages 2 and 158.)

Erdős, Paul, and Szekeres, George. 1960. On some extremum problems in elementary geometry. *Ann. Univ. Sci. Budapest. Eötvös Sect. Math.*, **3–4**, 53–62. (Cited on pages 2 and 106.)

Erickson, Jeff, and Seidel, Raimund. 1995. Better lower bounds on detecting affine and spherical degeneracies. *Discrete Comput. Geom.*, **13**(1), 41–57. (Cited on page 59.)

Estivill-Castro, Vladimir, Heednacram, Apichat, and Suraweera, Francis. 2009. Reduction rules deliver efficient FPT-algorithms for covering points with lines. *ACM J. Exp. Algorithmics*, **14**, A1.7:1–26. (Cited on page 62.)

Euler, Leonhard. 1862. Fragmenta arithmetica ex Adversariis mathematicis depromta, C: Analysis Diophantea. Pages 204–263 of: *Opera postuma*, vol. I. Petropolis: Eggers. Theorem 65, p. 229. (Cited on page 144.)

Fáry, István. 1948. On straight-line representation of planar graphs. *Acta Sci. Math. (Szeged)*, **11**, 229–233. (Cited on page 150.)

Fekete, Sándor P., Lübbecke, Marco E., and Meijer, Henk. 2008. Minimizing the stabbing number of matchings, trees, and triangulations. *Discrete Comput. Geom.*, **40**(4), 595–621. (Cited on page 196.)

Flammenkamp, Achim. 1992. Progress in the no-three-in-line problem. *J. Combin. Theory Ser. A*, **60**(2), 305–311. (Cited on page 73.)

Flammenkamp, Achim. 1998. Progress in the no-three-in-line problem. II. *J. Combin. Theory Ser. A*, **81**(1), 108–113. (Cited on page 73.)

Flum, Jörg, and Grohe, Martin. 2006. *Parameterized Complexity Theory*. Texts in Theoretical Computer Science: An EATCS Series. Berlin: Springer. (Cited on pages 46, 47, and 49.)

Fortnow, Lance. 2013. *The Golden Ticket: P, NP, and the Search for the Impossible*. Princeton, NJ: Princeton University Press. (Cited on page 43.)

Freimer, Robert Wilson. 2000. *Investigations in Geometric Subdivisions: Linear Shattering and Cartographic Map Coloring*. Ph.D. thesis, Cornell University. (Cited on page 197.)

Fulek, Radoslav, and Tóth, Csaba D. 2015. Universal point sets for planar three-trees. *J. Discrete Algorithms*, **30**, 101–112. (Cited on page 151.)

Füredi, Zoltán. 1991. Maximal independent subsets in Steiner systems and in planar sets. *SIAM J. Discrete Math.*, **4**(2), 196–199. (Cited on pages 80, 88, and 91.)

Furstenberg, Hillel, and Katznelson, Yitzhak. 1989. A density version of the Hales-Jewett theorem for $k = 3$. *Discrete Math.*, **75**(1–3), 227–241. (Cited on page 91.)

Gajentaan, Anka, and Overmars, Mark H. 1995. On a class of $O(n^2)$ problems in computational geometry. *Comput. Geom. Th. & Appl.*, **5**(3), 165–185. (Cited on pages 59 and 60.)

Garey, Michael R., and Johnson, David S. 1979. *Computers and Intractibility: A Guide to the Theory of NP-Completeness.* New York: W. H. Freeman. (Cited on page 96.)

Gil, Joseph, Steiger, William, and Wigderson, Avi. 1992. Geometric medians. *Discrete Math.*, **108**(1–3), 37–51. (Cited on page 140.)

Giménez, Omer, and Noy, Marc. 2009. Asymptotic enumeration and limit laws of planar graphs. *J. Amer. Math. Soc.*, **22**(2), 309–329. (Cited on page 183.)

Goldreich, Oded. 2010. *Property Testing: Current Research and Surveys.* Lecture Notes in Computer Science, vol. 6390. Berlin: Springer. (Cited on page 38.)

Goodman, Jacob E., and Pollack, Richard. 1986. Upper bounds for configurations and polytopes in $\mathbf{R}^d$. *Discrete Comput. Geom.*, **1**(3), 219–227. (Cited on pages 6 and 10.)

Goodman, Jacob E., Pollack, Richard, and Sturmfels, Bernd. 1989. Coordinate representation of order types requires exponential storage. Pages 405–410 of: *21st Symposium on Theory of Computing (STOC '89).* New York: Association for Computing Machinery. (Cited on page 33.)

Grantson, Magdalene, and Levcopoulos, Christos. 2006. Covering a set of points with a minimum number of lines. Pages 6–17 of: *6th Italian Conf. on Algorithms and Complexity (CIAC 2006).* Lecture Notes in Computer Science, vol. 3998. Berlin: Springer. (Cited on page 62.)

Green, Ben, and Tao, Terence. 2013. On sets defining few ordinary lines. *Discrete Comput. Geom.*, **50**(2), 409–468. (Cited on page 55.)

Grønlund, Allan, and Pettie, Seth. 2014. Threesomes, degenerates, and love triangles. Pages 621–630 of: *55th Symposium on Foundations of Computer Science (FOCS '14).* Los Alamitos, CA: IEEE Computer Society. (Cited on page 60.)

Grünbaum, Branko. 2003. *Convex polytopes.* 2nd edn. Graduate Texts in Mathematics, vol. 221. Berlin: Springer. See in particular Grünbaum's discussion of the Perles configuration on pp. 93–94. (Cited on pages 12, 143, and 144.)

Guibas, Leonidas J., Overmars, Mark H., and Robert, Jean-Marc. 1996. The exact fitting problem in higher dimensions. *Comput. Geom. Th. & Appl.*, **6**(4), 215–230. (Cited on page 61.)

Guillemot, Sylvain, and Marx, Dániel. 2014. Finding small patterns in permutations in linear time. Pages 82–101 of: *25th ACM–SIAM Symposium on Discrete Algorithms (SODA '14).* Philadelphia: Society for Industrial and Applied Mathematics. (Cited on page 165.)

Guth, Larry, and Katz, Nets Hawk. 2015. On the Erdős distinct distances problem in the plane. *Ann. of Math. (2)*, **181**(1), 155–190. (Cited on page 7.)

Guy, Richard K., and Kelly, Patrick A. 1968. The no-three-in-line problem. *Canad. Math. Bull.*, **11**, 527–531. (Cited on page 73.)

Hales, Alfred W., and Jewett, Robert I. 1963. Regularity and positional games. *Trans. Amer. Math. Soc.*, **106**, 222–229. (Cited on page 90.)

Hall, Richard R., Jackson, Terence H., Sudbery, Anthony, and Wild, Ken. 1975. Some advances in the no-three-in-line problem. *J. Combin. Theory Ser. A*, **18**, 336–341. (Cited on pages 74 and 75.)

Har-Peled, Sariel. 2009. *Approximating spanning trees with low crossing number.* Electronic preprint arxiv:0907.1131. (Cited on page 196.)

Har-Peled, Sariel, and Jones, Mitchell. 2017. *On separating points by lines.* Electronic preprint arxiv:1706.02004. (Cited on pages 192 and 197.)

Har-Peled, Sariel, and Lidický, Bernard. 2013. Peeling the grid. *SIAM J. Discrete Math.*, **27**(2), 650–655. (Cited on page 130.)

Harborth, Heiko. 1998. Integral distances in point sets. Pages 213–224 of: *Charlemagne and His Heritage: 1200 Years of Civilization and Science in Europe, Vol. 2 (Aachen, 1995).* Turnhout, Belgium: Brepols. (Cited on pages 144 and 145.)

Hartnett, Kevin. 2017. A puzzle of clever connections nears a happy end. *Quanta,* www.quantamagazine.org/a-puzzle-of-clever-connections-nears-a-happy-end-20170530/. (Cited on page 2.)

Haussler, David, and Welzl, Emo. 1987. ϵ-nets and simplex range queries. *Discrete Comput. Geom.*, **2**(2), 127–151. (Cited on page 70.)

Hearn, Robert A., and Demaine, Erik D. 2009. *Games, Puzzles, and Computation.* Wellesley, MA: A K Peters. (Cited on page 44.)

Hesse, Otto. 1844. Über die Elimination der Variabeln aus drei algebraischen Gleichungen vom zweiten Grade mit zwei Variabeln. *J. Reine Angew. Math.*, **28**, 68–96. (Cited on page 92.)

Higman, Graham. 1952. Ordering by divisibility in abstract algebras. *Proc. London Math. Soc., 3rd Ser.*, **2**, 326–336. (Cited on page 23.)

Huttenlocher, Daniel P., and Kedem, Klara. 1990. Computing the minimum Hausdorff distance for point sets under translation. Pages 340–349 of: *6th Symposium on Computational Geometry (SoCG '90).* New York: Association for Computing Machinery. (Cited on page 36.)

Impagliazzo, Russell, Paturi, Ramamohan, and Zane, Francis. 2001. Which problems have strongly exponential complexity? *J. Comput. System Sci.*, **63**(4), 512–530. (Cited on page 48.)

Jackson, John. 1821. Trees planted in rows. Pages 33–34 of: John Jackson, *Rational Amusement for Winter Evenings, or, A Collection of Above 200 Curious and Interesting Puzzles and Paradoxes.* London: n.p. (Cited on pages 53 and 125.)

Jadhav, Shreesh, and Mukhopadhyay, Asish. 1994. Computing a centerpoint of a finite planar set of points in linear time. *Discrete Comput. Geom.*, **12**(3), 291–312. (Cited on page 137.)

Jarník, Vojtěch. 1926. Über die Gitterpunkte auf konvexen Kurven. *Math. Z.*, **24**(1), 500–518. (Cited on page 110.)

Kalbfleisch, James G., and Stanton, Ralph G. 1995. On the maximum number of coplanar points containing no convex n-gons. *Utilitas Math.*, **47**, 235–245. (Cited on page 106.)

Károlyi, Gyula, and Solymosi, József. 2006. Erdős–Szekeres theorem with forbidden order types. *J. Combin. Theory Ser. A*, **113**(3), 455–465. (Cited on page 110.)

Károlyi, Gyula, and Tóth, Géza. 2012. Erdős–Szekeres theorem for point sets with forbidden subconfigurations. *Discrete Comput. Geom.*, **48**(2), 441–452. (Cited on pages 20 and 110.)

Károlyi, Gyula, Pach, János, and Tóth, Géza. 2001. A modular version of the Erdős–Szekeres theorem. *Studia Sci. Math. Hungar.*, **38**, 245–259. (Cited on page 135.)

Kelly, Leroy M., and Moser, William O. J. 1958. On the number of ordinary lines determined by n points. *Canad. J. Math.*, **10**, 210–219. (Cited on page 55.)

Kemnitz, Arnfried, and Harborth, Heiko. 2001. Plane integral drawings of planar graphs. *Discrete Math.*, **236**(1–3), 191–195. (Cited on page 148.)

Khuller, Samir, and Mitchell, Joseph S. B. 1990. On a triangle counting problem. *Inform. Process. Lett.*, **33**(6), 319–321. (Cited on page 140.)

Kitaev, Sergey. 2011. *Patterns in Permutations and Words*. EATCS Monographs in Theoretical Computer Science. Berlin: Springer. (Cited on page 156.)

Kleber, Michael. 2008. Encounter at far point. *Mathematical Intelligencer*, **1**, 50–53. (Cited on pages 147 and 148.)

Klee, Victor, and Wagon, Stan. 1991. Approximation by rational sets. Pages 54–57, 132–136 of: *Old and New Unsolved Problems in Plane Geometry and Number Theory*. Cambridge: Cambridge University Press. (Cited on page 142.)

Kløve, Torleiv. 1978. On the no-three-in-line problem. II. *J. Combin. Theory Ser. A*, **24**(1), 126–127. (Cited on page 73.)

Kløve, Torleiv. 1979. On the no-three-in-line problem. III. *J. Combin. Theory Ser. A*, **26**(1), 82–83. (Cited on page 73.)

Knuth, Donald E. 1992. *Axioms and Hulls*. Lecture Notes in Computer Science, vol. 606. Berlin: Springer. (Cited on pages 11 and 34.)

Komusiewicz, Christian, and Niedermeier, Rolf. 2012. New races in parameterized algorithmics. Pages 19–30 of: *37th International Symposium on Mathematical Foundations of Computer Science (MFCS 2012)*. Lecture Notes in Computer Science, vol. 7464. Berlin: Springer. (Cited on page 29.)

Kratsch, Stefan, Philip, Geevarghese, and Ray, Saurabh. 2016. Point line cover: The easy kernel is essentially tight. *ACM Trans. Algorithms*, **12**(3), A40:1–16. (Cited on page 62.)

Kreisel, Tobias, and Kurz, Sascha. 2008. There are integral heptagons, no three points on a line, no four on a circle. *Discrete Comput. Geom.*, **39**(4), 786–790. (Cited on pages 147 and 148.)

Kruskal, Joseph B. 1972. The theory of well-quasi-ordering: A frequently discovered concept. *J. Combin. Theory Ser. A*, **13**, 297–305. (Cited on page 22.)

Kurowski, Maciej. 2004. A 1.235 lower bound on the number of points needed to draw all n-vertex planar graphs. *Inform. Process. Lett.*, **92**(2), 95–98. (Cited on page 181.)

Lagrange, Jean, and Leech, John. 1986. Two triads of squares. *Math. Comp.*, **46**(174), 751–758. (Cited on page 146.)

Langerman, Stefan, and Morin, Pat. 2005. Covering things with things. *Discrete Comput. Geom.*, **33**(4), 717–729. (Cited on page 61.)

Lefmann, Hanno. 2012. *Extensions of the no-three-in-line problem*. Revision of a paper from Proc. AAIM 2008, www.tu-chemnitz.de/informatik/ThIS/downloads/publications/lefmann_no_three_submitted.pdf. (Cited on page 80.)

Liu, Regina Y. 1990. On a notion of data depth based on random simplices. *Ann. Statist.*, **18**(1), 405–414. (Cited on page 140.)

Lovász, László. 1979. *Combinatorial Problems and Exercises.* Amsterdam: North-Holland. (Cited on page 106.)

Loyd, Sam. 1914. Christopher Columbus shows some egg tricks. Page 301 of: *Sam Loyd's Cyclopedia of 5000 Puzzles, Tricks, and Conundrums with Answers.* New York: Lamb Publishing. (Cited on page 54.)

Marcus, Adam, and Tardos, Gábor. 2004. Excluded permutation matrices and the Stanley-Wilf conjecture. *J. Combin. Theory Ser. A*, **107**(1), 153–160. (Cited on page 6.)

Matiyasevich, Yuri V. 1993. *Hilbert's Tenth Problem.* Foundations of Computing Series. Cambridge, MA: MIT Press. (Cited on page 144.)

Matoušek, Jiří. 1991a. Computing the center of planar point sets. Pages 221–230 of: Goodman, Jacob E., Pollack, Richard, and Steiger, William L. (eds.), *Discrete and Computational Geometry: Papers from the DIMACS Special Year.* DIMACS Series in Discrete Mathematics and Theoretical Computer Science, vol. 6. Providence, RI: American Mathematical Society. (Cited on page 139.)

Matoušek, Jiří. 1991b. Spanning trees with low crossing number. *RAIRO Inform. Théor. Appl.*, **25**(2), 103–123. (Cited on page 191.)

Mazurkiewicz, Stefan. 1916. Sur un ensemble fermé, punctiforme, qui rencontre toute droite passant par un certain domaine. *Prace Mat.-Fiz.*, **27**, 11–16. (Cited on page 188.)

McMahon, Liz, Gordon, Gary, Gordon, Hannah, and Gordon, Rebecca. 2016. *The Joy of Set: The Many Mathematical Dimensions of a Simple Card Game.* Princeton, NJ: Princeton University Press. (Cited on page 91.)

Megiddo, Nimrod, and Tamir, Arie. 1982. On the complexity of locating linear facilities in the plane. *Oper. Res. Lett.*, **1**(5), 194–197. (Cited on page 61.)

Middendorf, Matthias, and Pfeiffer, Frank. 1992. The max clique problem in classes of string-graphs. *Discrete Math.*, **108**(1–3), 365–372. (Cited on page 151.)

Moon, John, and Moser, Leo. 1963. On Hamiltonian bipartite graphs. *Israel J. Math.*, **1**, 163–165. (Cited on page 101.)

Morris, Walter D., and Soltan, Valeriu. 2000. The Erdős–Szekeres problem on points in convex position: A survey. *Bull. Amer. Math. Soc.*, **37**(4), 437–458. (Cited on page 106.)

Moser, Robin A., and Tardos, Gábor. 2010. A constructive proof of the general Lovász local lemma. *J. ACM*, **57**(2), A11. (Cited on page 83.)

Mustafa, Nabil H., and Ray, Saurabh. 2014. Near-optimal generalisations of a theorem of Macbeath. Pages 578–589 of: *31st International Symposium on Theoretical Aspects of Computer Science (STACS '14).* Leibniz International Proceedings in Informatics (LIPIcs), vol. 25. Dagstuhl, Germany: Schloss Dagstuhl–Leibniz-Zentrum für Informatik. (Cited on page 71.)

OEIS Foundation. 2017. A003035: Maximal number of 3-tree rows in n-tree orchard problem. In: *The On-line Encyclopedia of Integer Sequences,* http://oeis.org/A003035. (Cited on pages 54 and 55.)

Pach, János, and Sharir, Micha. 2009. Sylvester–Gallai problem: The beginnings of combinatorial geometry. Pages 1–12 of: *Combinatorial Geometry and Its Algorithmic Applications: The Alcalá Lectures*. Mathematical Surveys and Monographs, vol. 152. Providence, RI: American Mathematical Society. (Cited on page 53.)

Payne, Michael S., and Wood, David R. 2013. On the general position subset selection problem. *SIAM J. Discrete Math.*, **27**(4), 1727–1733. (Cited on pages 80, 83, and 84.)

Peeples, William D. Jr. 1954. Elliptic curves and rational distance sets. *Proc. Amer. Math. Soc.*, **5**, 29–33. (Cited on page 146.)

Pegg, Ed. Jr. 2005 (April 11). Math Games: Chessboard tasks, www.mathpuzzle.com/MAA/36-Chessboard%20Tasks/mathgames_04_11_05.html. (Cited on page 74.)

Pilz, Alexander, and Welzl, Emo. 2015. Order on order types. Pages 285–299 of: Arge, Lars, and Pach, János (eds.), *31st International Symposium on Computational Geometry (SoCG '15)*. Leibniz International Proceedings in Informatics (LIPIcs), vol. 34. Dagstuhl, Germany: Schloss Dagstuhl–Leibniz-Zentrum für Informatik. (Cited on page 19.)

Roth, Klaus F. 1951. On a problem of Heilbronn. *J. London Math. Soc.*, **26**, 198–204. (Cited on page 74.)

Saaty, Thomas L., and Kainen, Paul C. 1986. *The Four-Color Problem: Assaults and Conquest.* 2nd edn. Mineola, NY: Dover. (Cited on page 45.)

Schaefer, Marcus, and Umans, Christopher. 2002. Complexity theory column 37: Completeness in the polynomial-time hierarchy: A compendium. *SIGACT News*, **33**(3), 32–49. Updated 2008 on the authors' web site, http://ovid.cs.depaul.edu/documents/phcom.pdf. (Cited on pages 44 and 171.)

Schnyder, Walter. 1990. Embedding planar graphs on the grid. Pages 138–148 of: *1st ACM/SIAM Symposium on Discrete Algorithms (SODA '90)*. Philadelphia: Society for Industrial and Applied Mathematics. (Cited on pages 181 and 182.)

Shearer, James B. 1985. On a problem of Spencer. *Combinatorica*, **5**(3), 241–245. (Cited on page 81.)

Shelah, Saharon. 1988. Primitive recursive bounds for van der Waerden numbers. *J. Amer. Math. Soc.*, **1**(3), 683–697. (Cited on page 90.)

Shneiderman, Ben, and Mayer, Richard. 1979. Syntactic/semantic interactions in programmer behavior: A model and experimental results. *International Journal of Computer & Information Sciences*, **8**(3), 219–238. (Cited on page 13.)

Shor, Peter W. 1991. Stretchability of pseudolines is NP-hard. Pages 531–554 of: Gritzmann, Peter, and Sturmfels, Bernd (eds.), *Applied Geometry and Discrete Mathematics: The Victor Klee Festschrift.* DIMACS Series in Discrete Mathematics and Theoretical Computer Science, vol. 4. Providence, RI: American Mathematical Society. (Cited on pages 44 and 177.)

Solymosi, József, and de Zeeuw, Frank. 2010. On a question of Erdős and Ulam. *Discrete Comput. Geom.*, **43**(2), 393–401. (Cited on page 149.)

Spencer, Joel. 1977. Asymptotic lower bounds for Ramsey functions. *Discrete Math.*, **20**(1), 69–76. (Cited on page 81.)

Sprinzak, Josef, and Werman, Michael. 1994. Affine point matching. *Pattern Recognition Letters*, **15**(4), 337–339. (Cited on page 36.)

Stein, Sherman K. 1951. Convex maps. *Proc. Amer. Math. Soc.*, **2**(3), 464–466. (Cited on page 150.)

Steinitz, Ernst. 1913. Bedingt konvergente Reihen und konvexe Systeme. *J. Reine Angew. Math.*, **143**(143), 128–175. (Cited on page 126.)

Suk, Andrew. 2017. On the Erdős–Szekeres convex polygon problem. *J. Amer. Math. Soc.*, **30**, 1047–1053. (Cited on page 2.)

Székely, László A. 1997. Crossing numbers and hard Erdős problems in discrete geometry. *Combin. Probab. Comput.*, **6**(3), 353–358. (Cited on page 84.)

Szemerédi, Endre, and Trotter, William T. Jr. 1983. Extremal problems in discrete geometry. *Combinatorica*, **3**(3–4), 381–392. (Cited on page 84.)

Tóth, Gábor. 2001. Point sets with many k-sets. *Discrete Comput. Geom.*, **26**(2), 187–194. (Cited on page 18.)

Tukey, John W. 1975. Mathematics and the picturing of data. Pages 523–531 of: *Proceedings of the International Congress of Mathematicians (Vancouver, B.C., 1974)*, Vol. 2. (Cited on page 136.)

Urabe, Masatsugu. 1996. On a partition into convex polygons. *Discrete Appl. Math.*, **64**(2), 179–191. (Cited on page 131.)

Vapnik, Vladimir N., and Chervonenkis, Alexey Ya. 1971. On the uniform convergence of relative frequencies of events to their probabilities. *Theory Probab. Appl.*, **16**(2), 264–280. (Cited on page 70.)

Wagner, Klaus. 1936. Bemerkungen zum Vierfarbenproblem. *Jber. Deutsch. Math.-Verein.*, **46**, 26–32. (Cited on page 150.)

Wang, Jianxin, Li, Wenjun, and Chen, Jianer. 2010. A parameterized algorithm for the hyperplane-cover problem. *Theoret. Comput. Sci.*, **411**(44–46), 4005–4009. (Cited on page 62.)

Welzl, Emo. 1992. On spanning trees with low crossing numbers. Pages 233–249 of: Monien, Burkhard, and Ottmann, Thomas (eds.), *Data Structures and Efficient Algorithms: Final Report on the DFG Special Joint Initiative*. Lecture Notes in Computer Science, vol. 594. Berlin: Springer. (Cited on pages 191, 193, and 194.)

Wilson, Robin. 2014. *Four Colors Suffice: How the Map Problem Was Solved*. Princeton Science Library. Princeton, NJ: Princeton University Press. (Cited on page 45.)

Wood, David R. 2004. A note on colouring the plane grid. *Geombinatorics*, **13**(4), 193–196. (Cited on page 87.)

Ziegler, Günter M. 2008. Nonrational configurations, polytopes, and surfaces. *Mathematical Intelligencer*, **30**(3), 36–42. (Cited on page 143.)

Index